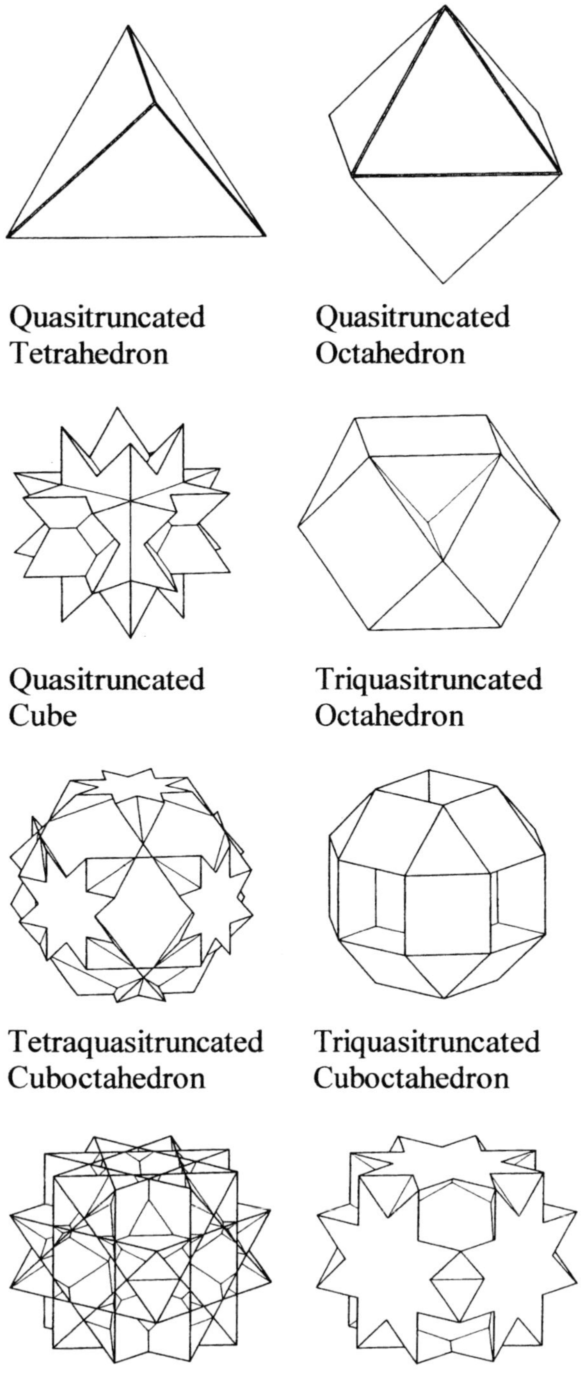

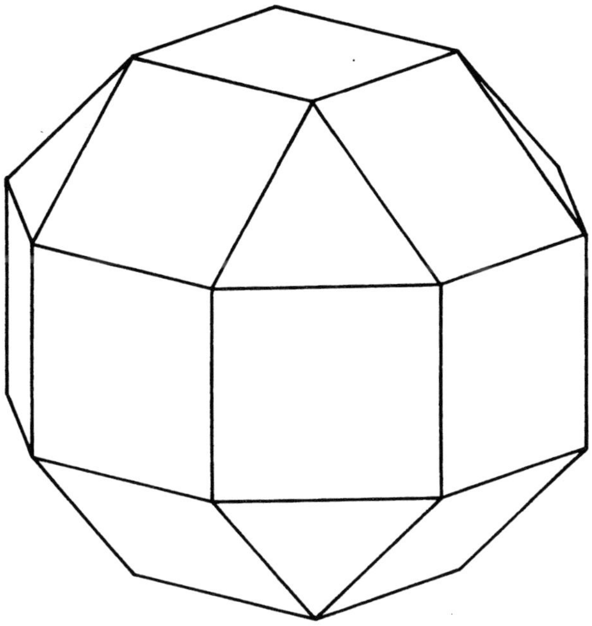

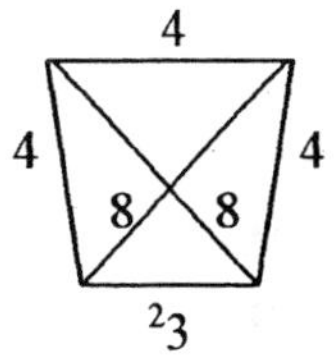

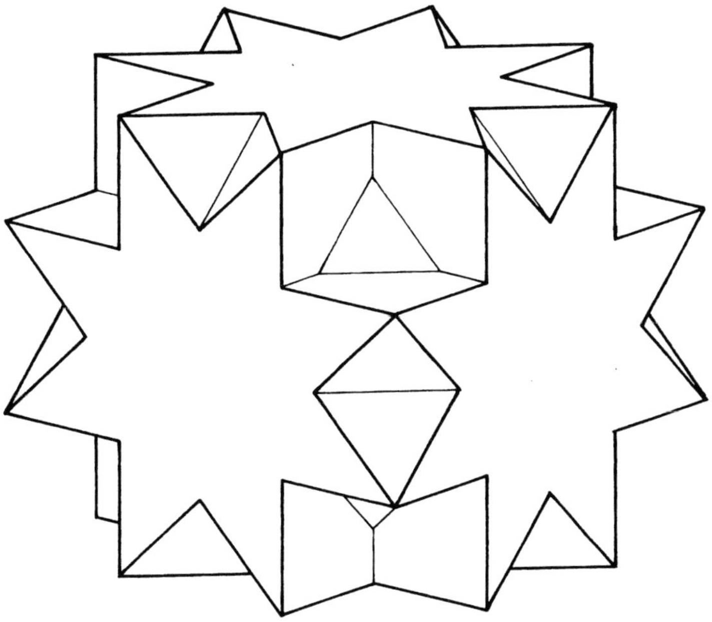

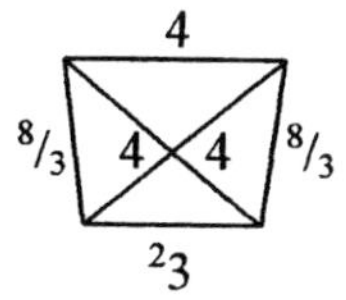

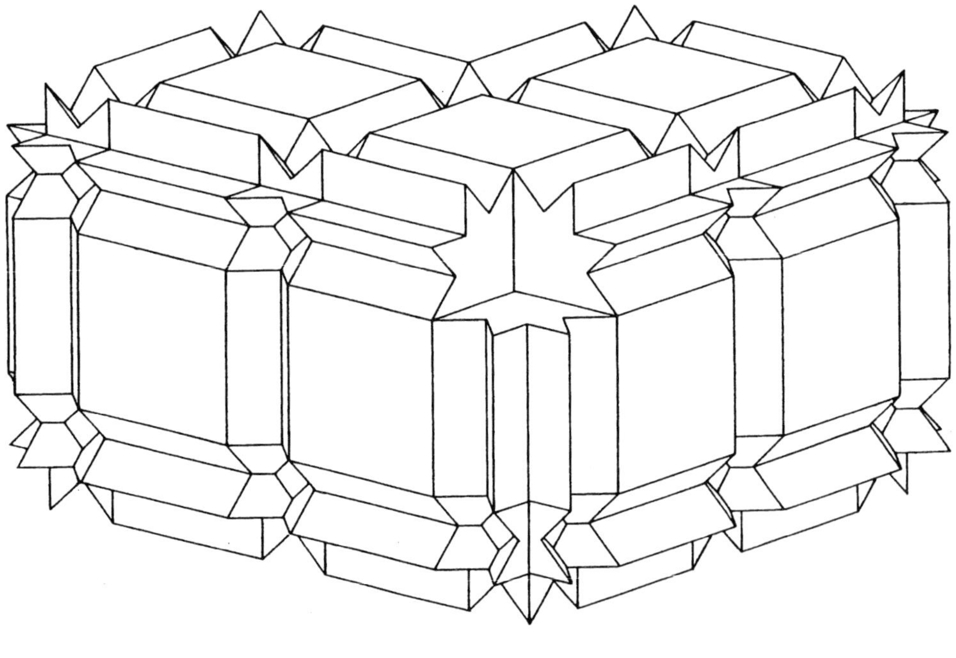

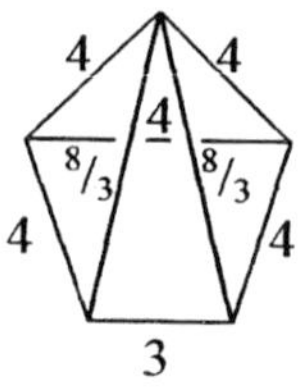

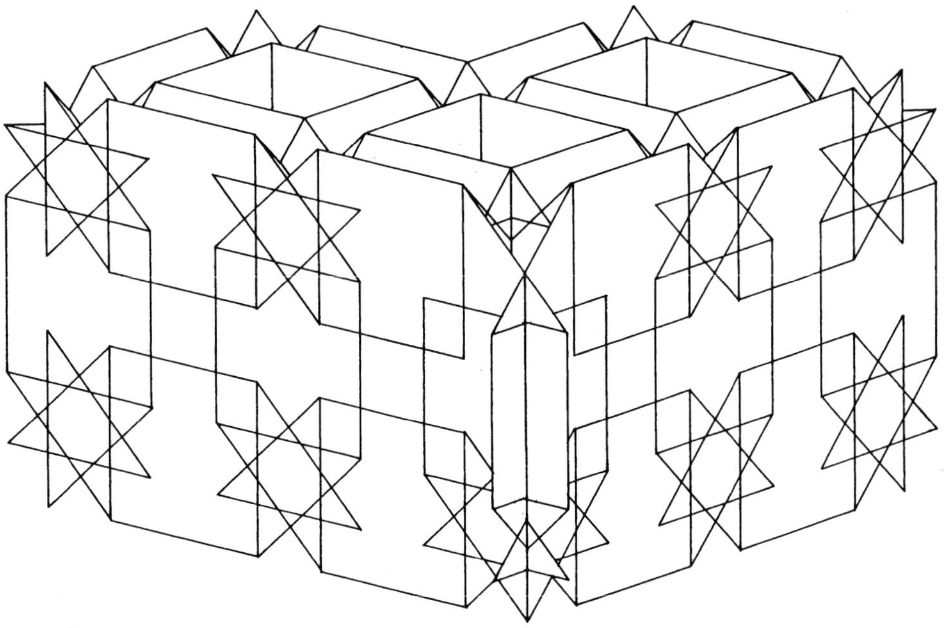

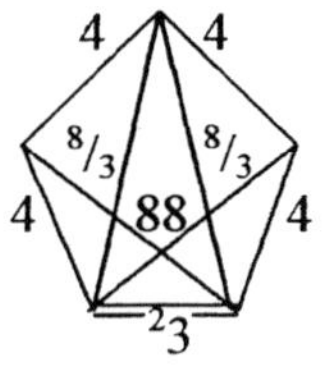

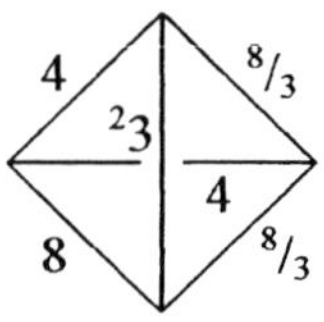

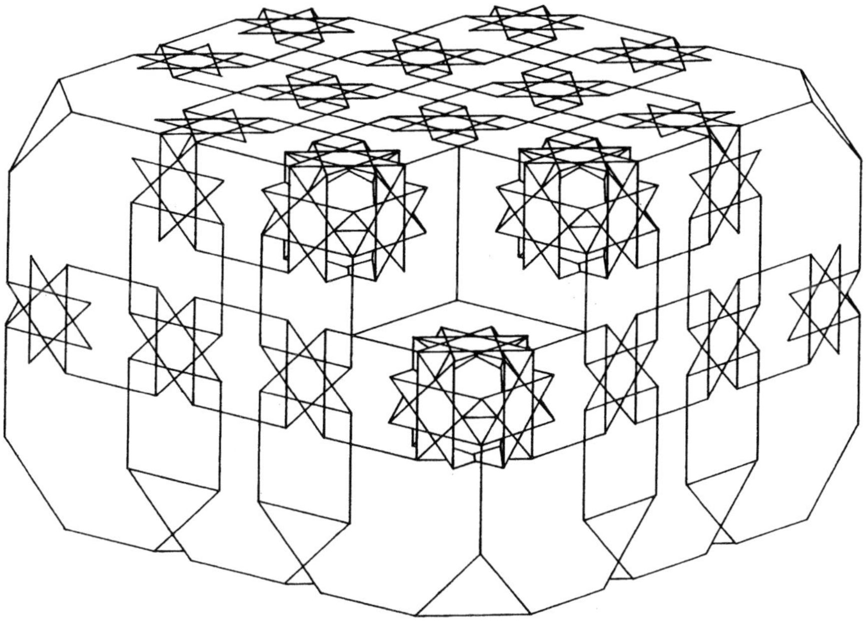

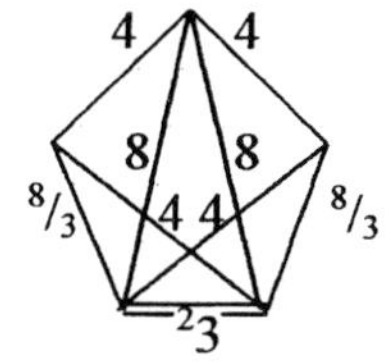

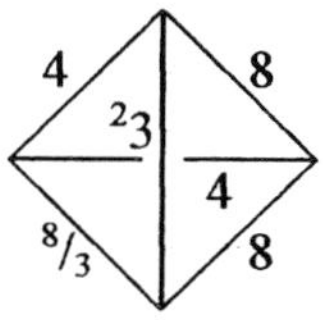

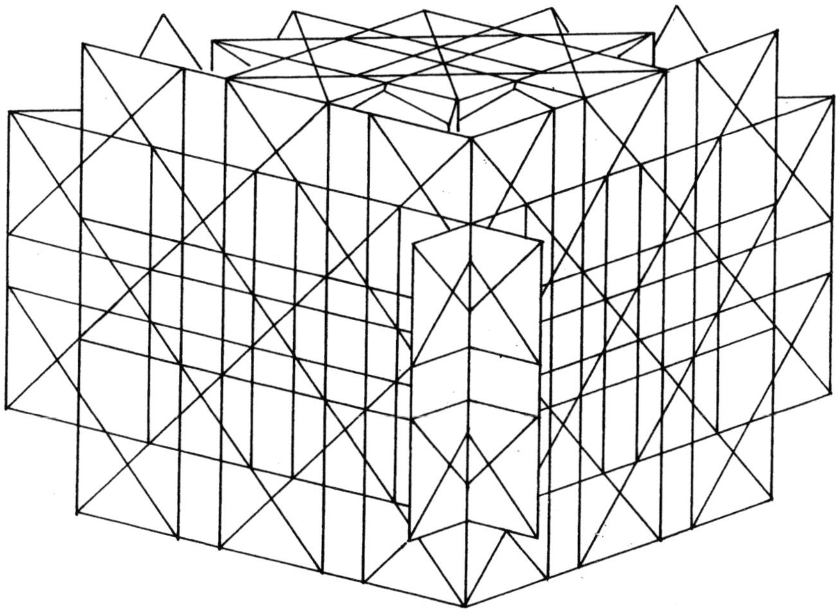

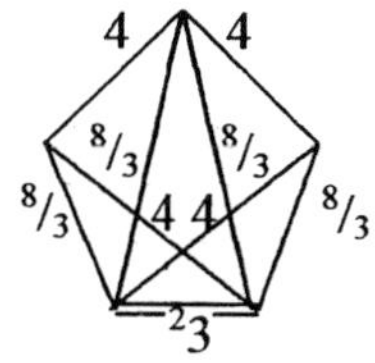

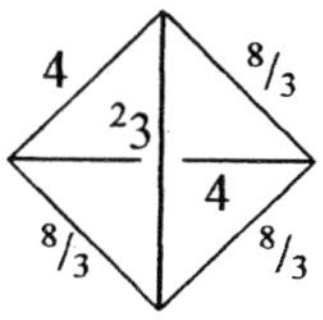

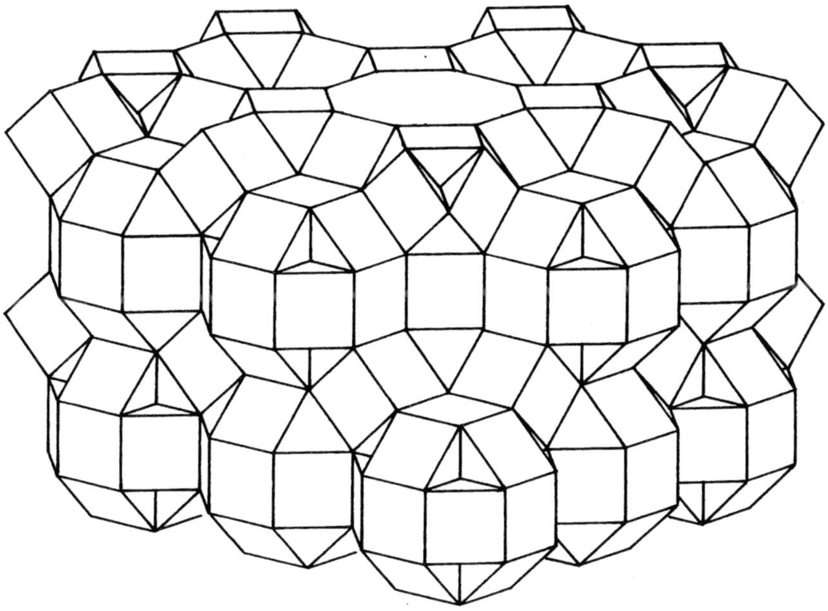

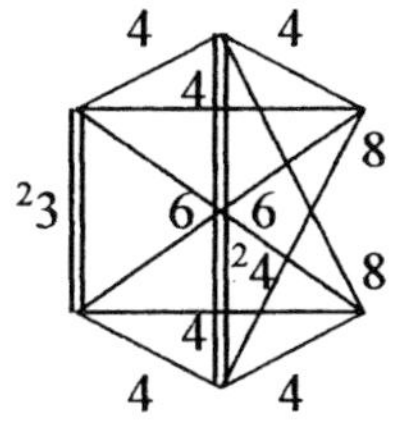

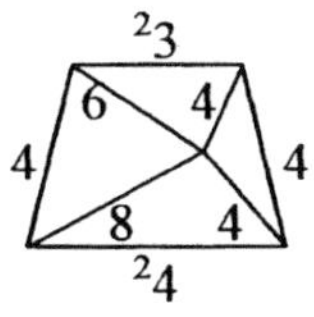

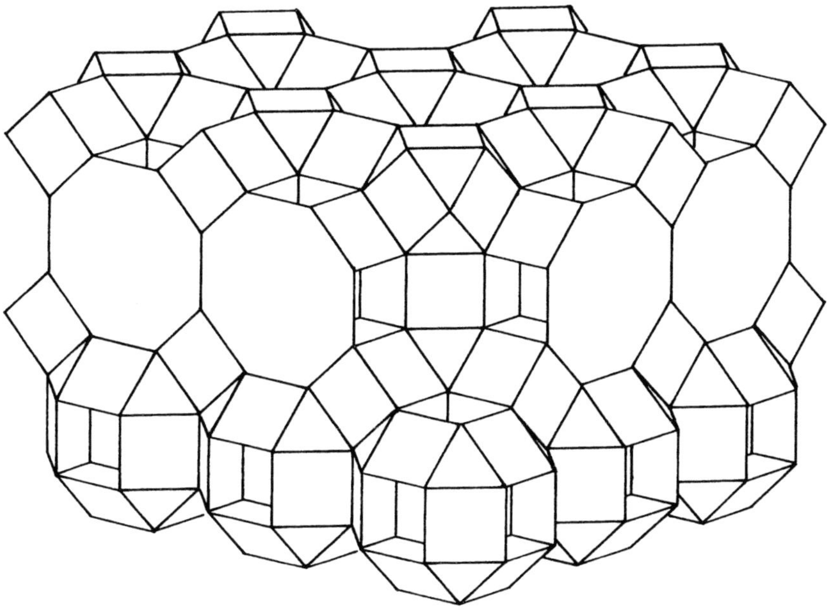

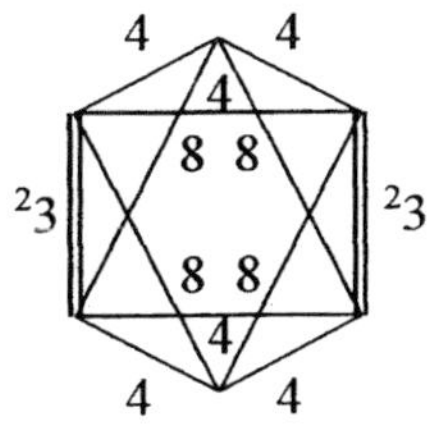

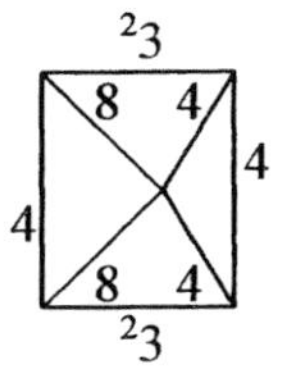

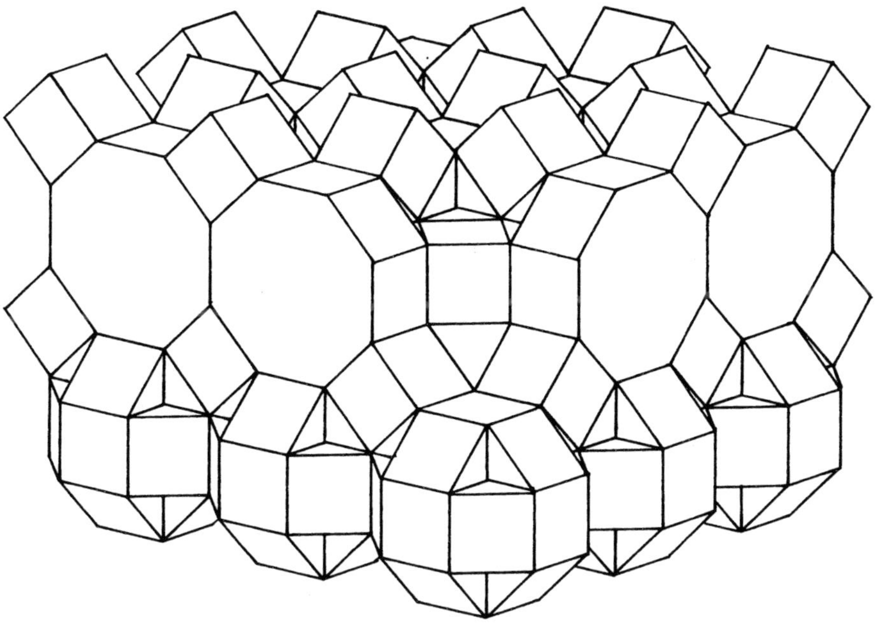

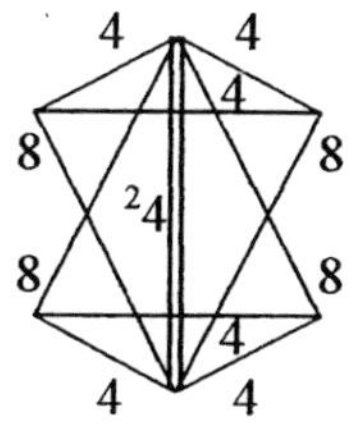

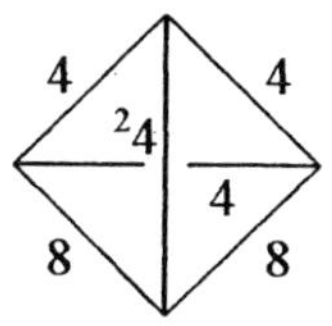

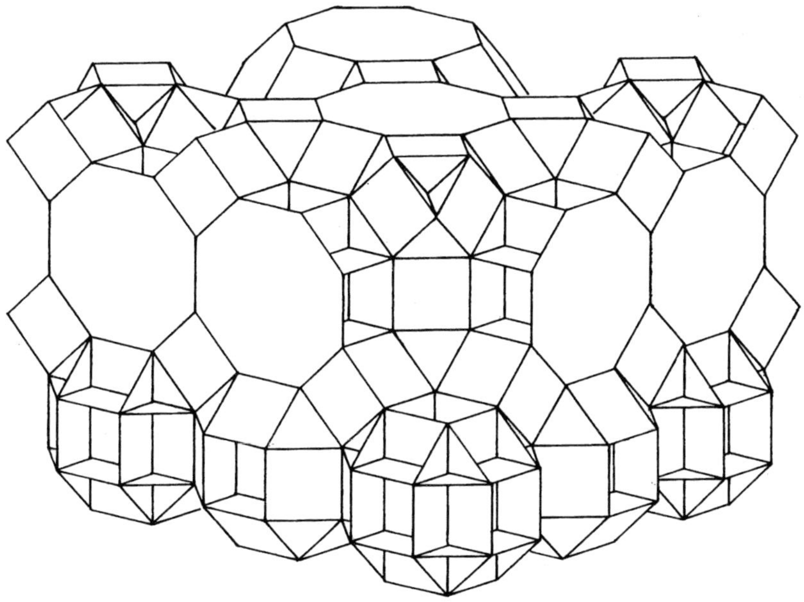

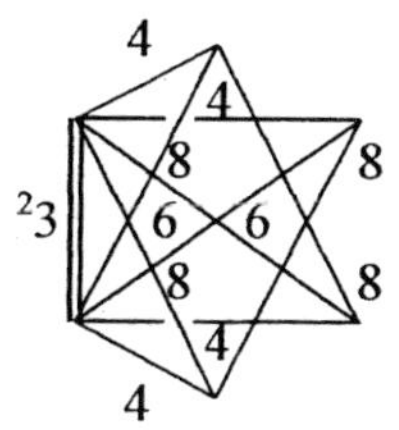

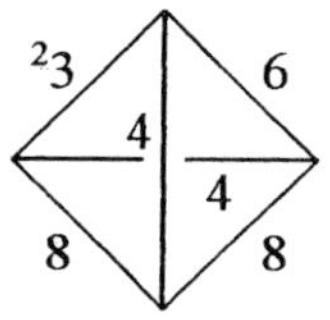

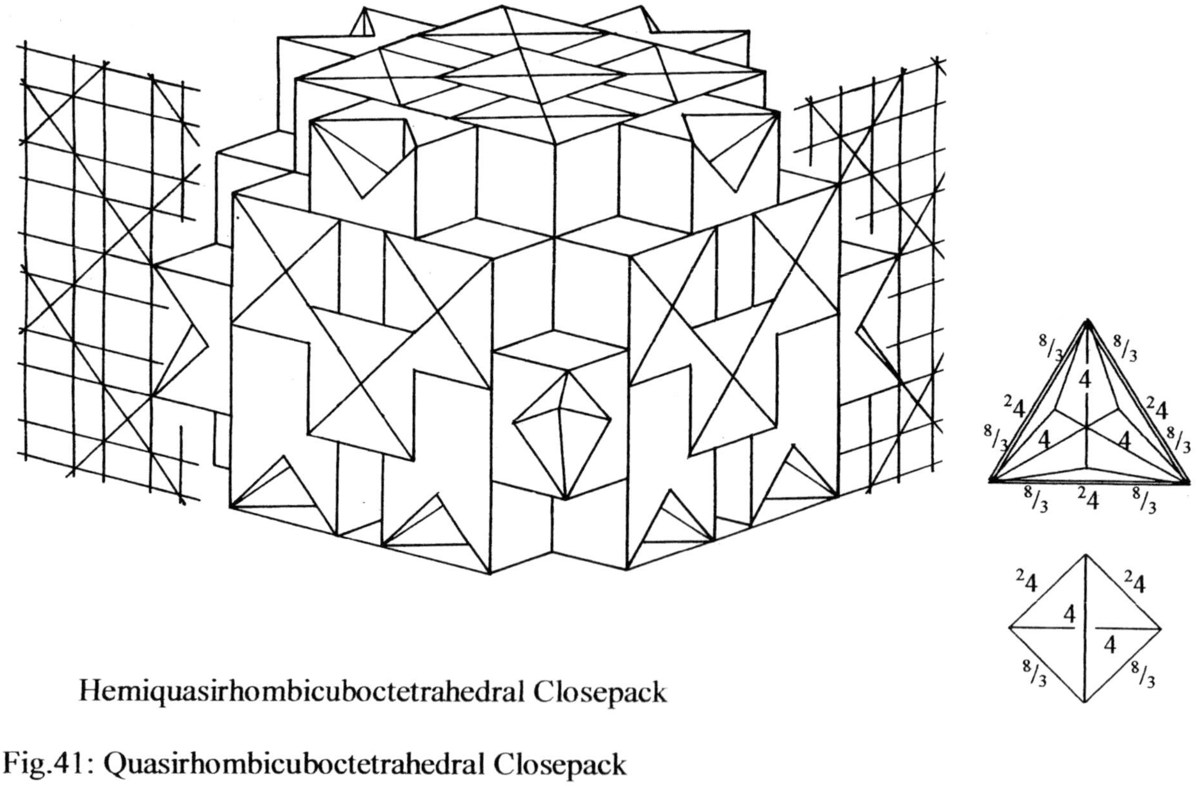

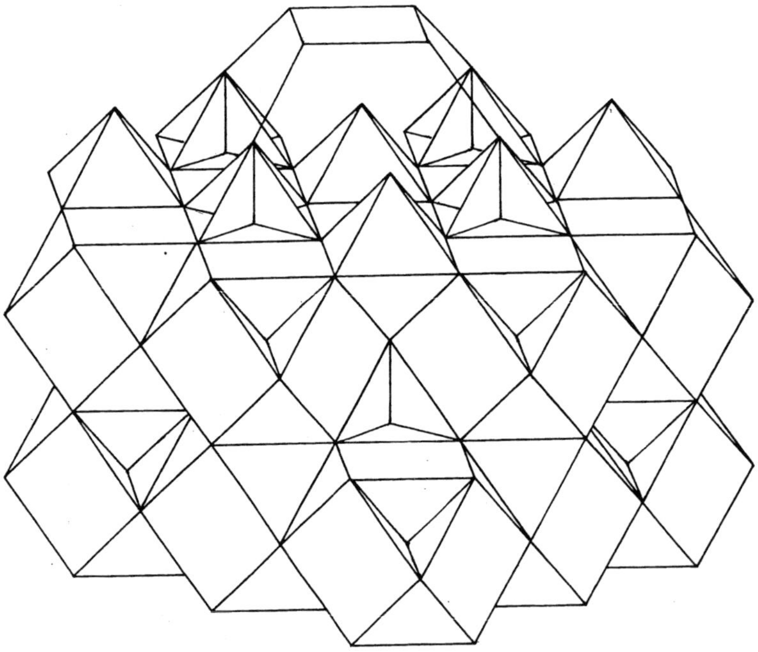

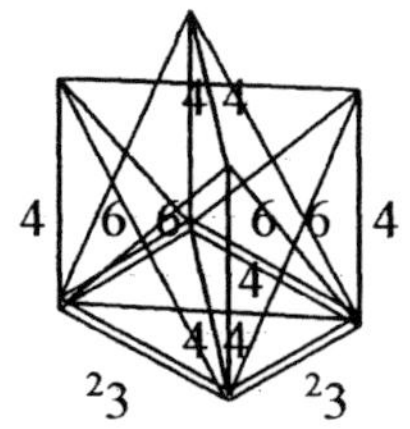

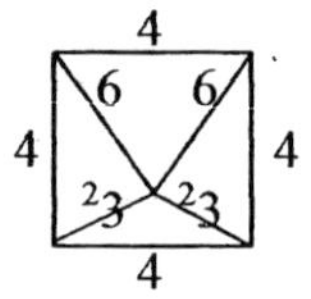

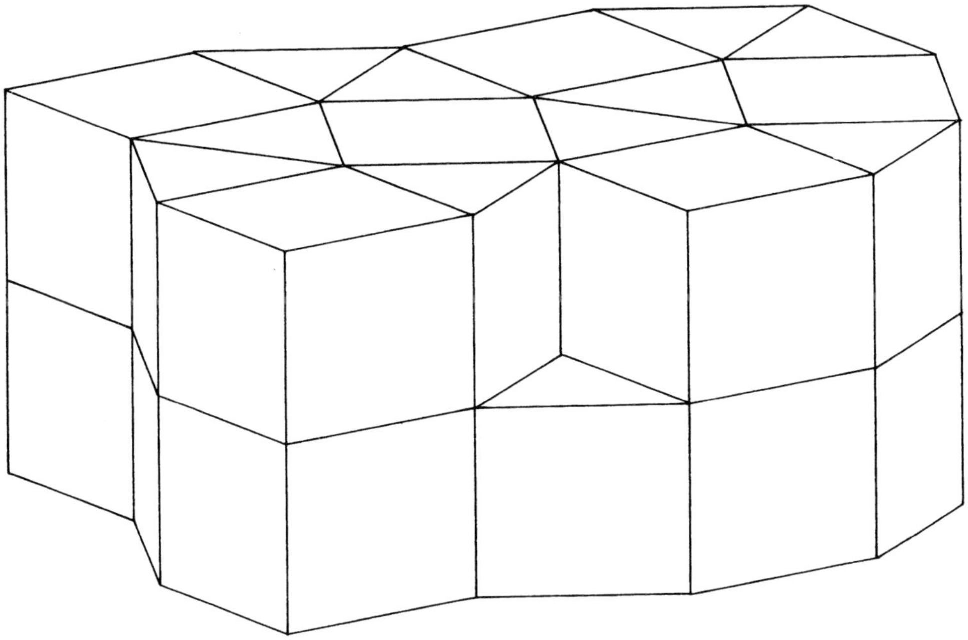

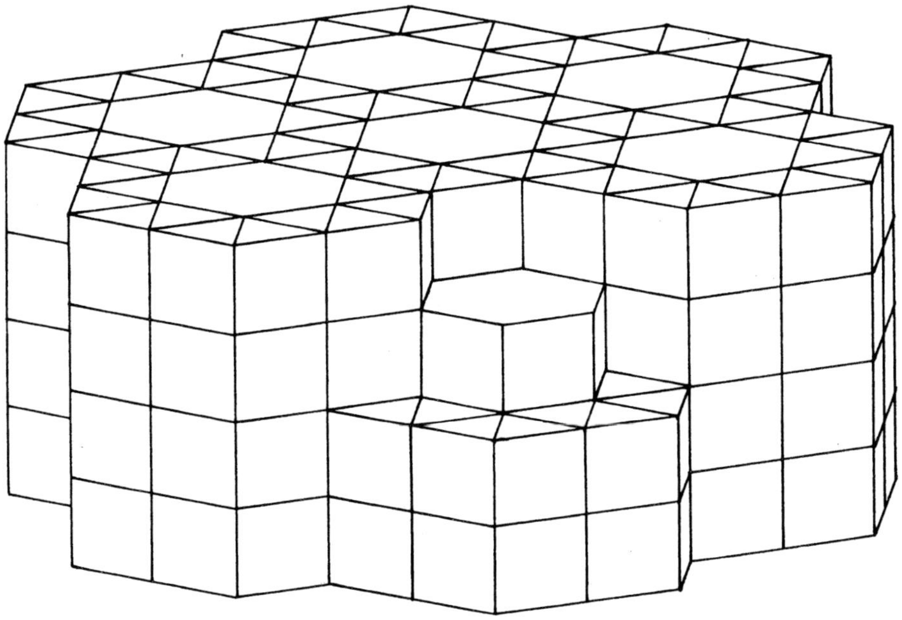

CLOSEPACKS & QUASI-CLOSEPACKS

BEING THE FIFTH PART
OF SEVERAL COMPRISING
THE COMPLETE? POLYHEDRA
BY PATRICK TAYLOR

CLOSEPACKS & QUASI-CLOSEPACKS

First published 2004 by

Nattygrafix
277 Cavendish Street
Ipswich
Suffolk
IP3 8BQ

e-mail: polystar@ntlworld.com

(01473) 434604

Made in England

ISBN 0 9516701 6 6

NATTYGRAFIX IPSWICH

CONTENTS

INTRODUCTION	5
'REGULAR' CLOSEPACKS	6
TRUNCATED CLOSEPACKS	8
EVEN FACED CLOSEPACKS	10
FURTHER TRUNCATIONS	12
ODD CLOSEPACKS	14
SUMMARY OF TRUE CLOSEPACKS	16
A RECAP ON QUASITRUNCATION	17
QUASITRUNCATED CLOSEPACKS	18
EVEN FACED QUASITRUNCATIONS	20
MORE EVEN FACED QUASITRUNCATIONS	22
A TRUE QUASIRHOMBI CLOSEPACK	24
2nd SERIES QUASITRUNCATIONS	26
ANOTHER QUASIRHOMBI CLOSEPACK	28
ODD QUASITRUNCATIONS	30
QUASIRHOMBIPRISM CLOSEPACKS	32
CLOSEPACKS WITH TILINGS	34
HEMITRUNCATED CLOSEPACKS	36
HEMIQUASITRUNCATED CLOSEPACKS	38
HEMIRHOMBIPRISM CLOSEPACKS	40
HEXAHEXAPRISM CLOSEPACKS	42
RHOMBICUBOCTETRAHEDRALS	44
RHOMBICUBOCTAHEDRALS	46
QUASIRHOMBICUBOCTETRAHEDRALS	48
QUASIRHOMBICUBOCTAHEDRALS	50
CUBOCTAHEDRAL TILING CLOSEPACKS	52
ANOTHER TRIPRISM VERTEX	54
ANOTHER CUBOID VERTEX	56
THE FINAL VERTEX SHELL	58
SUMMARY OF CLOSEPACKS	60
A NOTE ON REGULAR HONEYCOMBS	61
APPENDIX: TILING CLOSEPACKS	62
TRUNCATED TILING CLOSEPACKS	64
SNUB & ODD TILING CLOSEPACKS	66
MORE TRUNCATED TILING CLOSEPACKS	68
SO WHAT FOLLOWS?	70
BIBLIOGRAPHY	71

ILLUSTRATIONS

Figure		Page
1*	'Regular' Closepacks	7
2*	Rhombic Closepack	8
3*	Truncated Closepacks	9
4*	Second Truncation Closepack	10
5*	Third Truncation Closepacks	11
6*	Fourth Truncation Closepack	12
7*	Second Series Closepacks	13
8*	First Odd Closepack	14
9*	Other Odd Closepacks	15
10	Quasitruncated Polyhedra	17
11*	Second Quasitruncation	18
12*	Quasitruncated Closepacks	19
13*	Quasitruncated Closepack	20
14*	Quasitruncated Closepacks	21
15*	Quasitruncated Closepack	22
16*	Quasitruncated Closepacks	23
17*	Quasirhombi Closepack	24
18*	Quasitruncated Closepacks	25
19*	Quasitruncated Closepack	26
20*	Quasitruncated Closepacks	27
21*	Another Quasirhombi Closepack	28
22*	Quasitruncated Odd Closepacks	29
23*	Quasitruncated Odd Closepack	30
24*	Quasitruncated Odd Closepacks	31
25*	Quasirhombi Odd Closepack	32
26*	Quasirhombi Odd Closepacks	33
27*	Closepack with Tilings	34
28*	Closepacks with Tilings	35
29*	Hemirhombi Odd Closepack	36
30*	Hemitruncated Closepacks	37
31*	Hexahexaprism Closepack	38
32*	Hemiquasitruncated Closepacks	39
33*	Hemirhombi Odd Closepack	40

Figure		Page
34*	Hemiquasirhombi Odd Closepacks	41
35*	Hexahexaprism Closepack	42
36*	Quasihexahexaprism Closepacks	43
37*	Rhombicuboctetrahedral Cp	44
38*	Rhombicuboctahedral Closepacks	45
39*	Rhombicuboctahedral Closepack	46
40*	Rhombicuboctahedral Closepacks	47
41*	Quasirhombicuboctetrahedral Cp	48
42*	Quasirhombicuboctahedral Cps	49
43*	Quasirhombicuboctahedral Cp	50
44*	Quasirhombicuboctahedral Cps	51
45*	Cuboctahedral Tiling Closepack	52
46*	Cuboctahedral Tiling Closepacks	53
47*	Another Triprism Vertex Cp	54
48*	Double Cuboctaprism Closepacks	55
49*	Another Cuboid Vertex Closepack	56
50*	More Cuboid Vertex Closepacks	57
51	Assembling The Final Vertex	58
52	Within The Final Vertex	59
53	True, Quasi and Hemi Vertices	60
54	Regular Honeycombs	61
55*	Square Tiling Closepack	62
56*	Hexagon & Triangle Tiling Cps	63
57*	Quasi-Regular Tiling Closepack	64
58*	Truncated Tiling Closepacks	65
59*	Odd Tiling Closepack	66
60*	Snub Tiling Closepacks	67
61*	Rhombic Closepack	68
62*	More Truncated Tiling Closepacks	69
63	So What Follows?	70

Figures * are supported by Tables nos. 1-30

for Geoff Knight and Arthur Dodd (Ramblers and Mathematical Association)
crossed paths, shared interests and fresh impetus along the way

"Nowadays it is being realized that geometry is still a subject of abiding beauty that provides tremendous intellectual satisfaction in return for effort put into its study, ... Geometry is having a revival!"

Brannan, Esplen & Gray *Geometry* 1999

INTRODUCTION

The geometric shapes explored in this fifth volume of '*The Complete? Polyhedra*' are those formed by assembling polyhedra, themselves assemblies of polygons, into higher ordered forms that fill all of three dimensions. These will be found to fall into two main types. Firstly we have the true closepacks that fill three dimensions and shadow at a higher level the two dimensional tilings seen at the start of the second volume of this series '*Incomplete Tilings*'.

Secondly we have the quasi-closepacks, the higher analogue of the quasitruncated tilings. Whilst these do fill all of three dimensions (except for a few degenerate forms) they do so in a 'dense' way. The solids composing them are required to interpenetrate each other, much as the polygons in the quasi-tilings were required to overlap. The result is that any point in space will find itself within several different solids at once.

Having already explored and illustrated the polyhedra, we should find the true closepacks fairly simple to visualise, particularly as they are all based on the simpler tetrahedral {3,3} and octahedral {3,4} symmetry families of polyhedra.

Any difficulties encountered will probably arise with the quasi-closepacks, with their complex interpenetrations. Hopefully the rewards in terms of aesthetic beauty and completeness of the picture presented will make the effort of visualising these complex assemblies worthwhile. In the absence of any comprehensive naming system, other than that of naming each solid present, I have endeavoured to invent one that I think works and hopefully captures the essential quality of each closepack in turn.

‘REGULAR’ CLOSEPACKS

The best known and only truly regular closepack, an assembly of equal regular polyhedra that fills all of three dimensional space, is that of the cube, illustrated opposite in Figure 1.

Rather than the vertex-edge duality seen in polygons, or the vertex-face duality of the polyhedra, we find here vertex-solid duality, such that joining the centres of one such set of cubes, delineates another equal set of cubes. The cubic closepack is thus a self-dual and by joining together the triangular vertex figures of the eight cubes at any one vertex we can obtain the octahedral vertex figure of the closepack itself.

It will be remembered that the cube is in fact the rhombic solid of the {3,3} tetrahedral symmetry family, and interestingly we find that the rhombic dodecahedron, the equivalent solid of the {3,4} octahedral symmetry family, also forms a closepack that fills all of three dimensions. This is shown overleaf in Figure 2, but cannot be considered regular since the constituent rhombic dodecahedra are not themselves regular and the vertices are of two types.

Because of the two different types of vertex, one tetrahedral the other cubic, its dual consists of two types of solid closepacked together, octahedra and tetrahedra as shown opposite in Figure 1. Again this is not strictly a regular closepack, but might be considered quasi-regular by analogy with the polyhedra with two types of face. This has a cuboctahedral vertex figure.

In stick model form this closepack is the familiar ‘octet truss’, so beloved by engineers for large span roofs on account of its repetition of stable equilateral triangles.

Solids	E/V	F/V	S/V
cube	6	12 □	8
{4,3,4} = t^6{4,3,4}			
Cubic Closepack			
octahedron	12	24 △	6
tetrahedron			8
{octet}			
Octetrahedral Closepack			
rhombic dodecahedron	4	6 ◇	4
(two vertex types)	8	12	6
t^6{octet}			
Rhombic Dodecahedral Closepack			

Table 1: ‘Regular’ Closepacks

The properties of these three ‘regular’ closepacks are shown above in Table 1. Because they are infinite in extent like the tilings, we cannot compare total numbers of vertices, edges, faces or solids, but only their abundance relative to each vertex. The figures show edges per vertex (E/V), which equals the number of vertices in the vertex figure, faces per vertex (F/V), which totals to equal the number of edges in the vertex figure and finally solids per vertex (S/V) which totals to equal the number of faces in the vertex figure. Obviously every face will have two solids attached to it, and the number of solids around an edge will equal the number of faces around the same edge.

The Schläfli symbol for the cubic closepack is {4,3,4}, consisting of the {4,3} symbol of its constituent cubes, 4 of which are arranged around each edge. In general we will find that {x,y,z} consists of z {x,y} polyhedra arranged around each edge, its vertex figure is a {y,z} with {x} polygons for edges, and it will have a dual {z,y,x}.

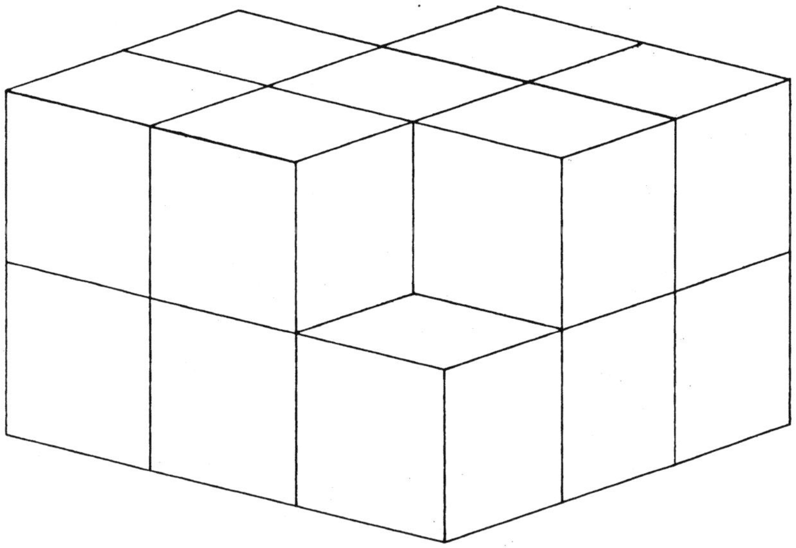

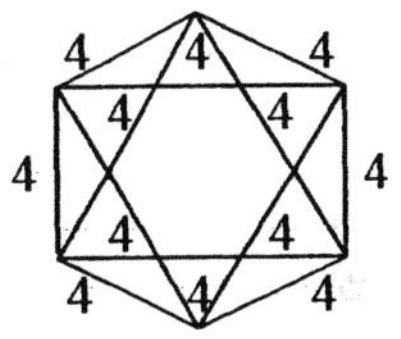

$\{4,3,4\} = t^6\{4,3,4\}$
Cubic Closepack

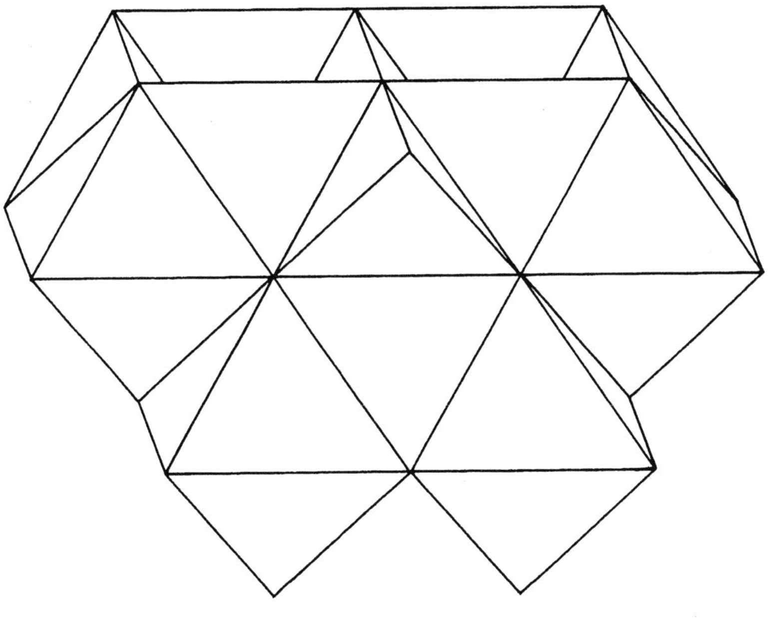

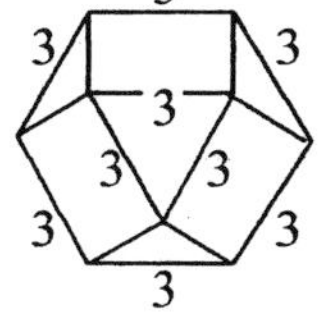

{octet}
Octetrahedral Closepack

Fig.1: ‘Regular’ Closepacks

TRUNCATED CLOSEPACKS

As with the tilings we can proceed from any regular closepack towards its dual by a series of 'truncations'. In this instance we grow a new solid (rather than face) at each vertex. The first semi-regular closepacks encountered in this process are shown opposite in Figure 3 and described adjoining in Table 2.

In each case the new 'grown' truncating solid is the same as the vertex figure of the original closepack, the remaining solids being truncated versions of those in the originals.

Both these truncated closepacks have pyramidal vertex figures, analogous to the isosceles triangle vertex figures we found with the first truncations of the tilings and polyhedra

Solids	E/V	F/V	S/V
truncated cube	5	4 ⬡	4
octahedron		4 △	1
t{4,3,4}			
Truncated Cubic Closepack			
truncated octahedron	5	4 ⬡	2
truncated tetrahedron		2 □	2
cuboctahedron		2 △	1
t{octet}			
Truncated Octetrahedral Closepack			
cuboctahedron	8	4 □	4
octahedron		8 △	2
$t^2\{4,3,4\} = t^2\{octet\}$			
Cuboctahedral Closepack			

Table 2: Truncated Closepacks

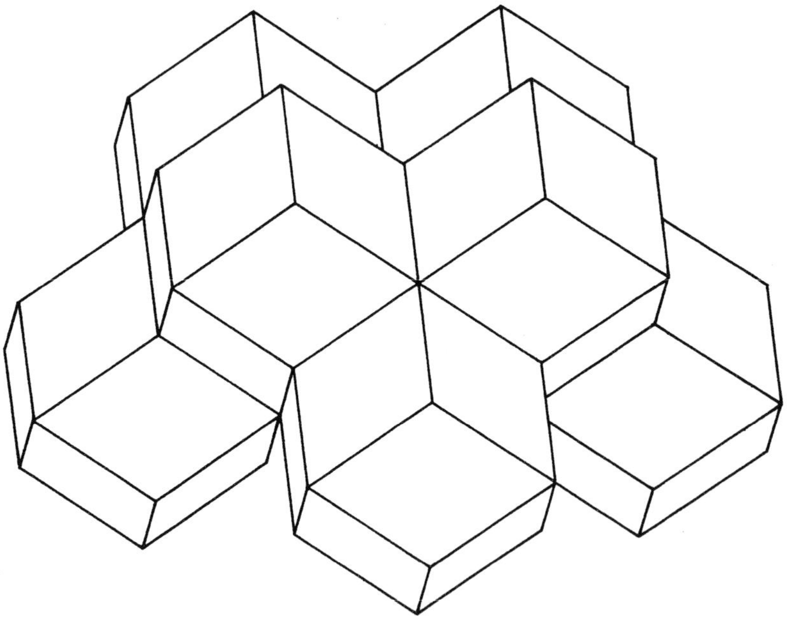

t^6\{octet\}
Rhombic Dodecahedral Closepack

Fig.2: Rhombic Closepack

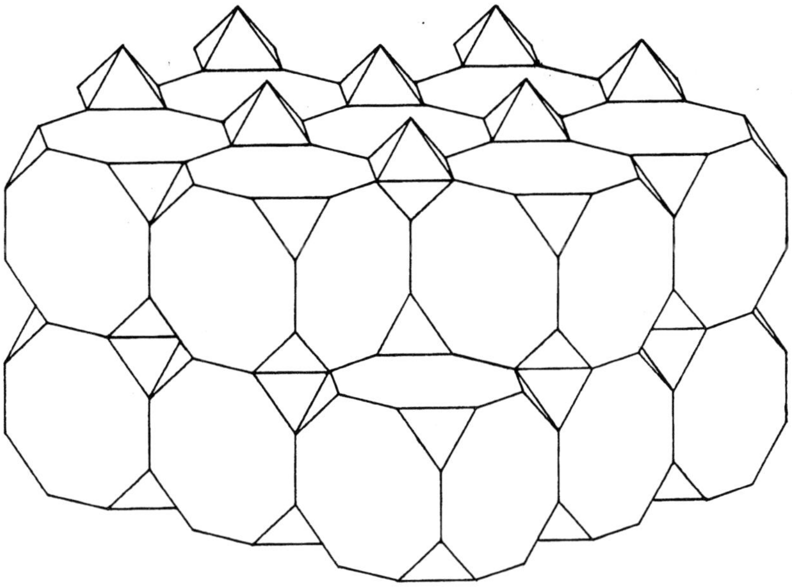

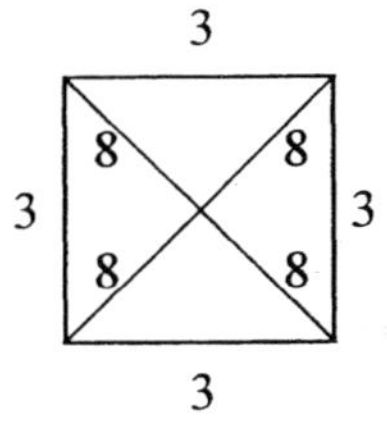

t{4,3,4}
Truncated Cubic Closepack

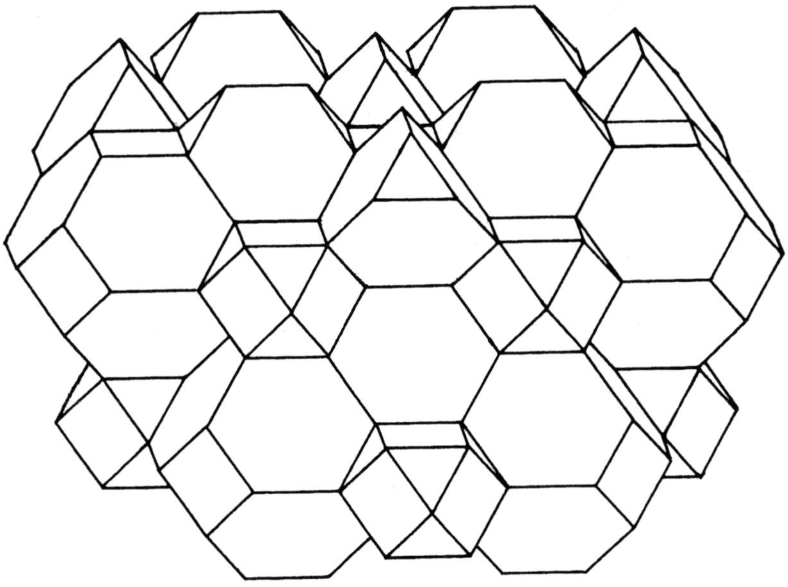

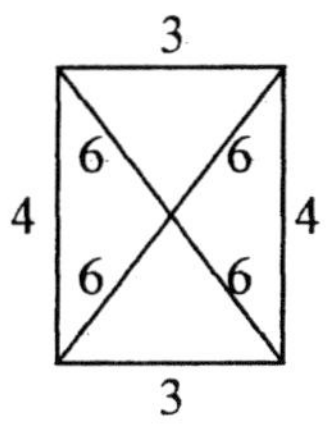

t{octet}
Truncated Octetrahedral Closepack

Fig.3: Truncated Closepacks

Surprisingly the second truncation in both series is the same closepack, that of cuboctahedra and octahedra shown below in Figure 4. This could be regarded as quasi-regular and has a cuboid vertex figure, there being four cuboctahedra and two octahedra at each vertex. It was shown as $t^2\{4,3,4\}$ in Table 2 overleaf and appears as the second truncation of the octetrahedral closepack in Table 3 adjoining.

Figure 5 opposite shows the next two truncations, the third truncation and central member in each of our two series. Since these consist entirely of truncated polyhedra of one kind or another by analogy with the even faced polyhedra (ignoring the triangles) they could be termed 'even faced' closepacks. They both have tetrahedral vertex figures.

Solids	E/V	F/V	S/V
truncated octahedron	4	2 □	2
truncated octahedron		4 ⬡	2
$t^3\{4,3,4\}$			
Truncated Octahedral Closepack			
truncated cube	4	1 △	1
truncated tetrahedron		1 □	1
truncated cuboctahedron		2 ⬡	2
$t^3\{octet\}$		2 ○	
Truncated Cuboctetrahedral Closepack			
cuboctahedron	8	4 △	2
octahedron		4 □	2
cuboctahedron		4 △	2
$t^2\{octet\} = t^2\{4,3,4\}$			
Cuboctetrahedral Closepack			

Table 3: More Truncated Closepacks

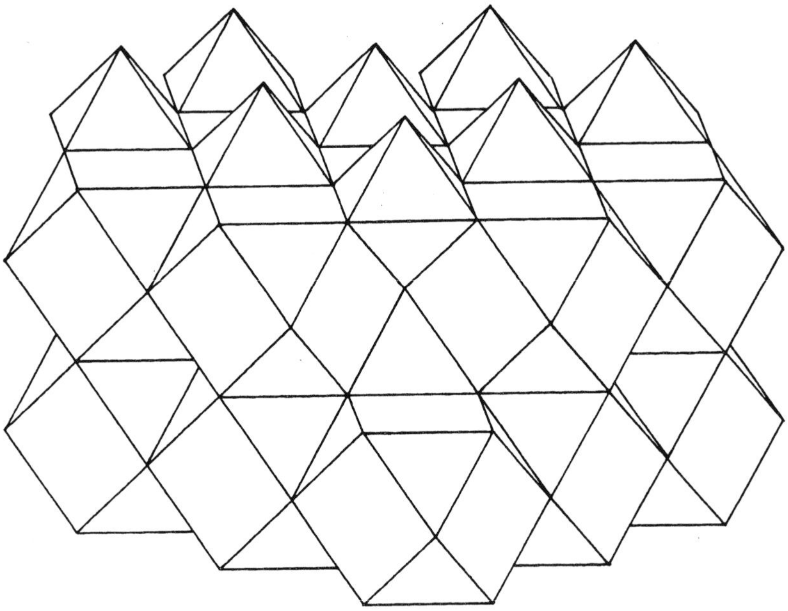

$t^2\{4,3,4\} = t^2\{octet\}$
Cuboctahedral Closepack

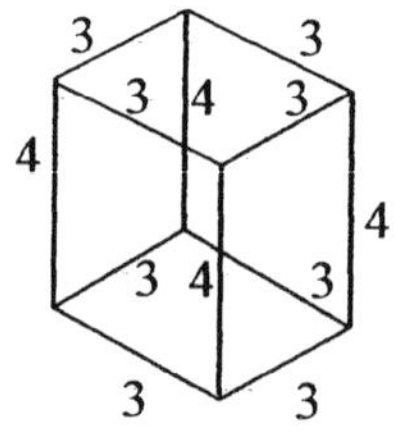

Fig.4: Second Truncation Closepack

'Quasi-Regular'

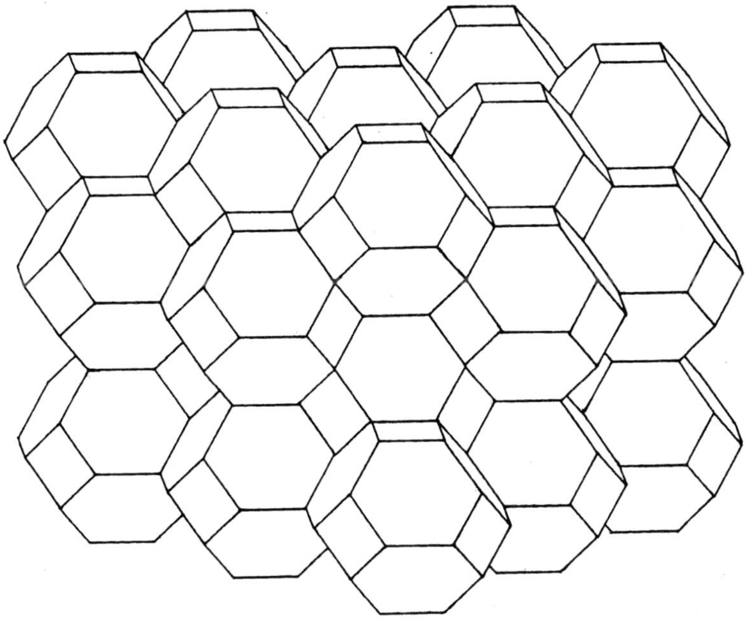

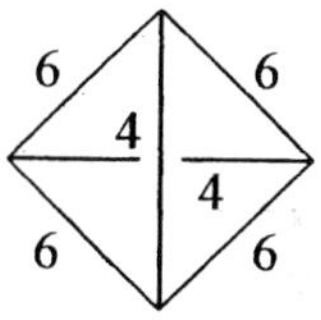

$t^3\{4,3,4\}$
Truncated Octahedral Closepack

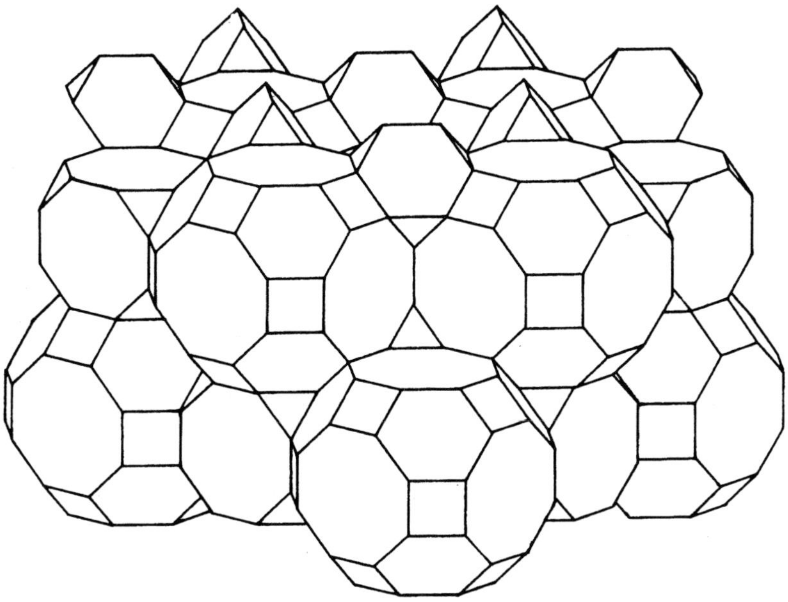

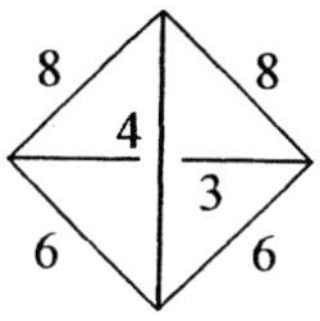

$t^3\{octet\}$
Truncated Cuboctetrahedral Closepack

Fig.5: Third Truncation Closepacks 'Even Faced'

FURTHER TRUNCATIONS

Having reached the centre of the self-dual {4,3,4} series, we will find that t^4 is the same as t^2{4,3,4} and that t^5 is the same as t{4,3,4}. However there is one further member of the octetrahedral closepack series, the fourth truncation shown below in Figure 6. It has a distorted triangular prism vertex figure and is the last in this series, as beyond lie only irregular forms and eventually the rhombic dodecahedral closepack (Figure 2).

Two further truncated closepacks are however possible, formed as part of a second series that starts with the cuboctahedral closepack These are shown in Figure 7 opposite, the first truncation having a tetrahedral vertex figure, the second another distorted triangular prism.

Solids	E/V	F/V	S/V
truncated cuboctahedron	4	3 □	2
truncated octahedron		2 ⬡	1
cube		1 ○	1
tt^2{4,3,4} = tt^2{octet}			
Truncated Cuboctahedral Closepack			
rhombicuboctahedron	6	2 △	2
cuboctahedron		7 □	1
cube			2
t^2t^2{4,3,4} = t^2t^2{octet}			
Rhombicuboctahedral Closepack			
cube	6	3 △	1
tetrahedron		6 □	1
rhombicuboctahedron			3
t^4{octet}			
Rhombicuboctetrahedral Closepack			

Table 4: Even More Truncated Closepacks

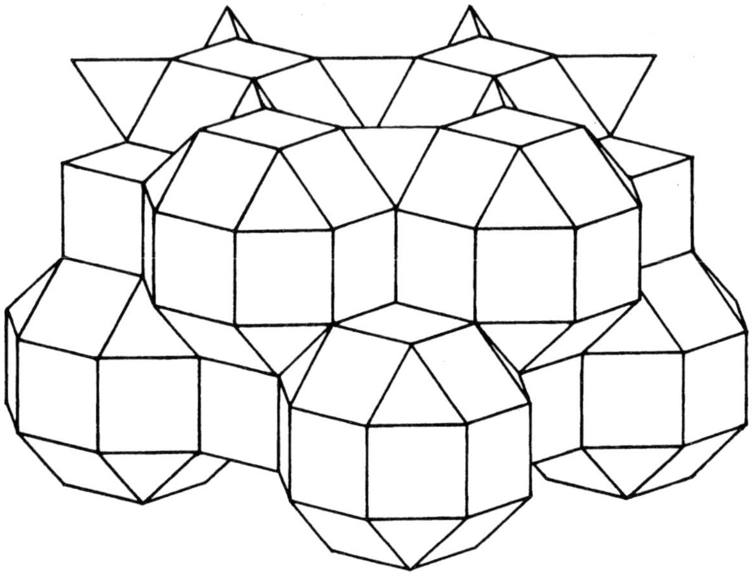

t^4{octet}
Rhombicuboctetrahedral Closepack

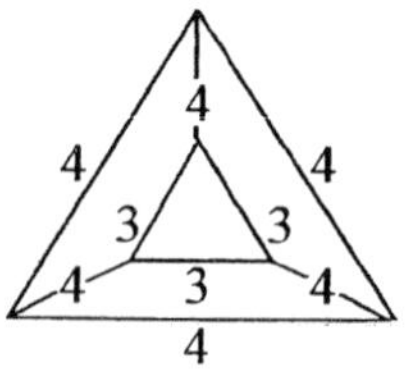

Fig.6: Fourth Truncation Closepack

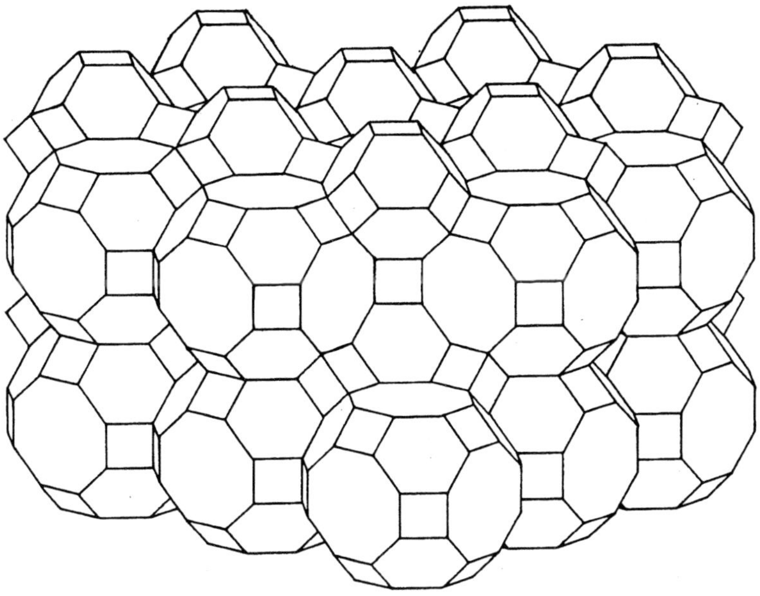

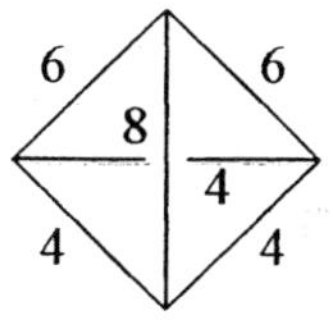

$tt^2\{4,3,4\} = tt^2\{octet\}$
Truncated Cuboctahedral Closepack

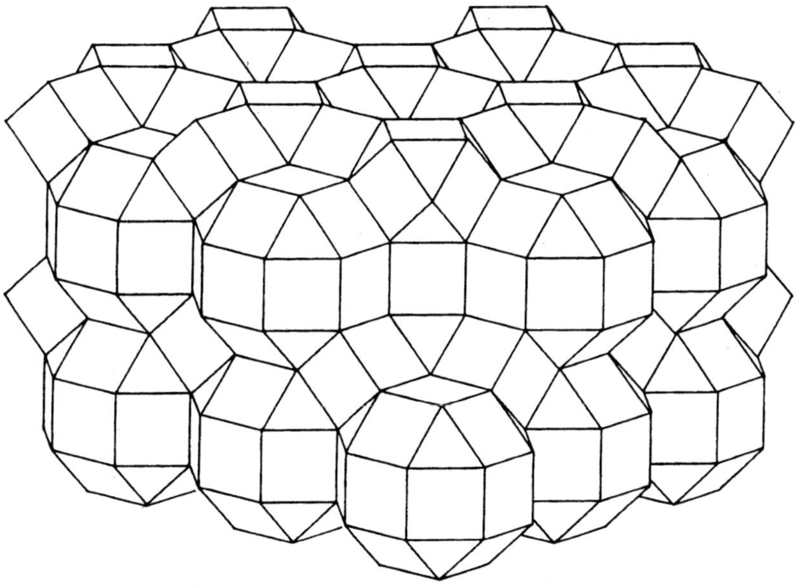

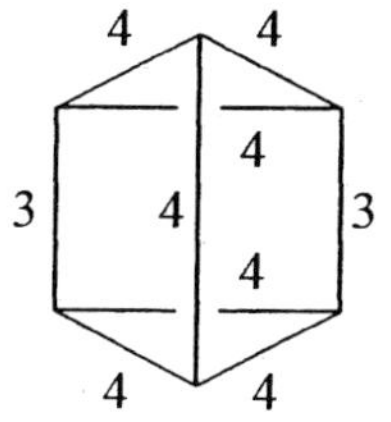

$t^2t^2\{4,3,4\} = t^2t^2\{octet\}$
Rhombicuboctahedral Closepack

Fig.7: Second Series Closepacks

ODD CLOSEPACKS

There remain three further closepacks that defy classification as part of a series of truncations, perhaps analogous to the odd tiling with two squares and three triangles at each vertex. These are described in Table 5 adjoining.

The truncated tetrahedral closepack shown below in Figure 8 has a stretched octahedral vertex figure, but apparently no dual or related closepacks.

The other two are shown opposite in Figure 9 and can be seen to be somewhat similar, both involving octagonal prisms in their make up. One has a tetrahedral vertex figure the other a pyramidal one, not unlike some of the truncated forms already seen.

Solids	E/V	F/V	S/V
	4	3 □	
truncated cuboctahedron		1 ⬡	2
octaprism		2 ○	2
Truncated Cuboctaprism Closepack			
rhombicuboctahedron	5	1 △	1
truncated cube		5 □	1
cube		2 ○	1
octaprism			2
Rhombicuboctaprism Closepack			
truncated tetrahedron	6	6 △	6
tetrahedron		6 ⬡	2
Truncated Tetrahedral Closepack			

Table 5: Odd Closepacks

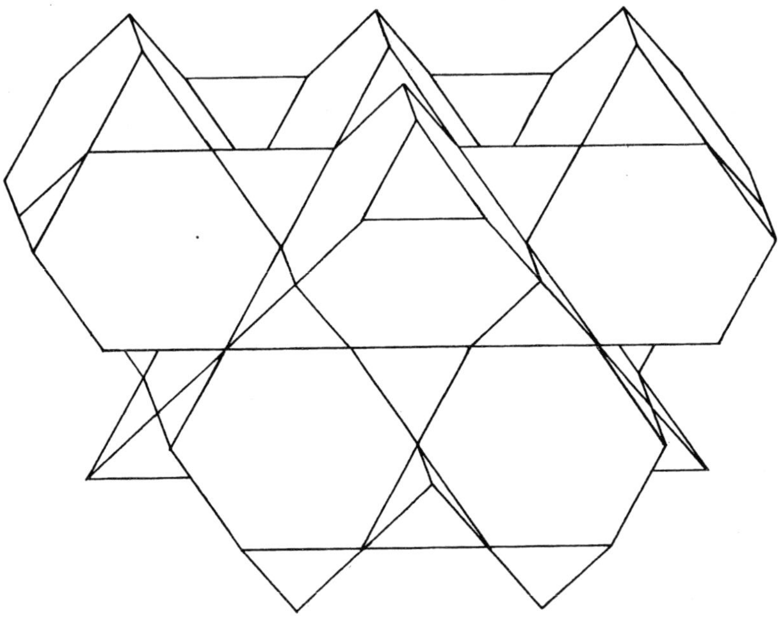

Truncated Tetrahedral Closepack

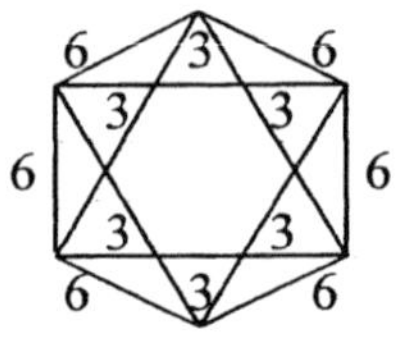

Fig.8: First Odd Closepack

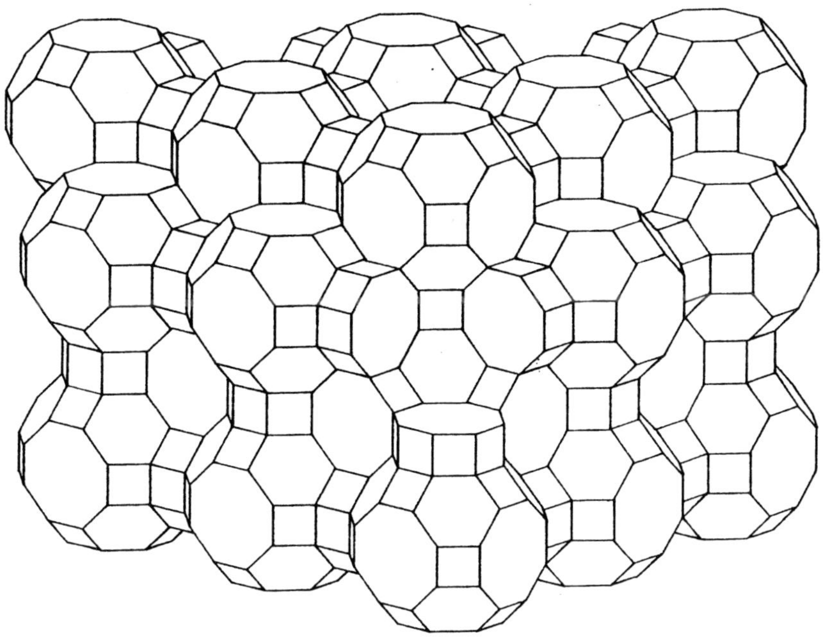

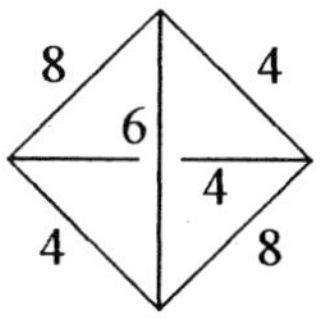

Truncated Cuboctaprism Closepack

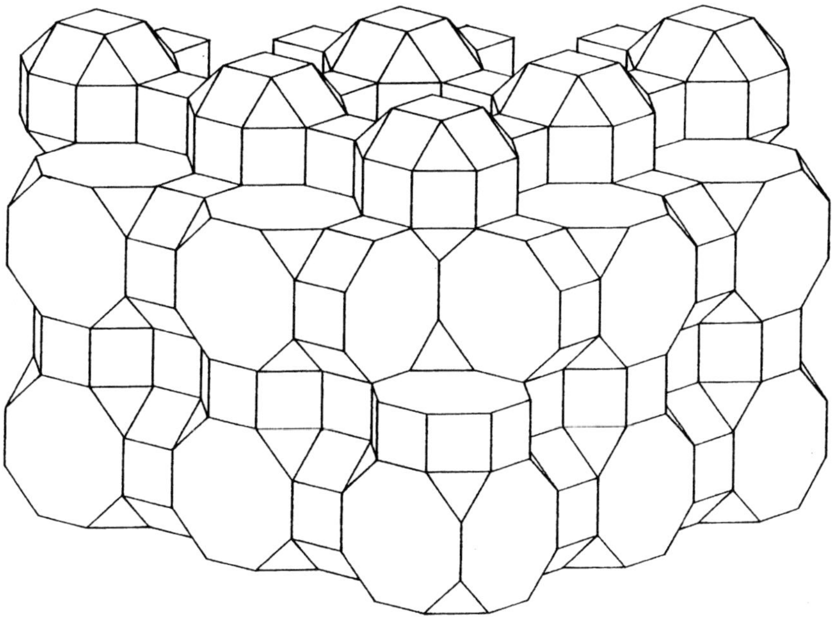

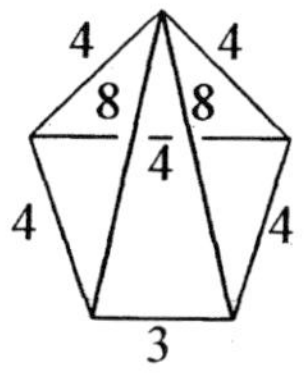

Rhombicuboctaprism Closepack

Fig.9: Other Odd Closepacks

SUMMARY OF TRUE CLOSEPACKS

cube	—
tr cube	octa
cubocta	octa
tr octa	tr octa
octa	cubocta
octa	tr cube
—	cube

{4,3,4} First Series

octa	tetra	—
tr octa	tr tetra	cubocta
cubocta	octa	cubocta
tr cube	tr tetra	tr cubocta
cube	tetra	rhombicubocta
		I R R E G U L A R
—	—	rhombic dodeca

{octet} First Series

cubocta	octa	—
tr cubocta	tr octa	cube
rhombicubocta	cubocta	cube

{4,3,4} Second Series

From what we have seen in the preceding pages, the fourteen 'true' closepacks seem to fall into series of truncations. These are not dissimilar to those seen for the tilings and polyhedra, which had series of truncations generally with five members, composed of two overlapping series of polygon truncations each of three members.

The closepacks in their turn, at a higher level of geometrical organisation, generally have series seven members long, composed of two overlapping polyhedral series as summarised in Table 6 adjoining.

The patterns we have seen here will also be found to hold for the hypersolids which we will examine in the next volume in this series.

tetra	tr tetra	
tr cubocta	octaprism	
rhombicubocta	octaprism	cube tr cube

Odd Closepacks

Table 6: Closepack Series

A RECAP ON QUASITRUNCATION

We found when dealing with the tilings that it was possible to quasitruncate them, thus introducing double and star polygons to our polygonal vocabulary, which necessitated overlapping faces and interpenetrating edges. In a similar way if we allow interpenetrating solids to occur, we can also produce quasitruncated closepacks, hence the term quasi-closepacks. This practice might seem to go against the grain in terms of strict closepacking, however inter-penetration will be found relatively normal, if not essential, when we come to consider the star hypersolids in due course.

Going back to basics, a quasitruncated polygon is one obtained by truncating the inverse of a polygon, rather than the polygon itself. With the polyhedra, quasitruncation can only be applied to the first truncation of a series, where truncated polygons are usually encountered. Quasitruncated polyhedra can thus be obtained by substituting quasitruncated for truncated polygons in these polyhedra.

Here with the closepacks, such truncated polyhedra can be found at first, third and fifth truncation point of each series, so the possibilities are fairly numerous. The pages that follow will examine these quasi-truncations along with a few 'quasirhombi' forms obtained by substituting quasi-rhombihedra for rhombihedra at the second and fourth truncation points of each series.

To assist in our visualisation of these complex forms, the quasitruncations of the polyhedra encountered so far in the true closepacks are illustrated adjoining in Figure 10. Many of these are deemed not 'uniform' in the standard classification, but their inclusion here will once again allow us to see a more complete? picture.

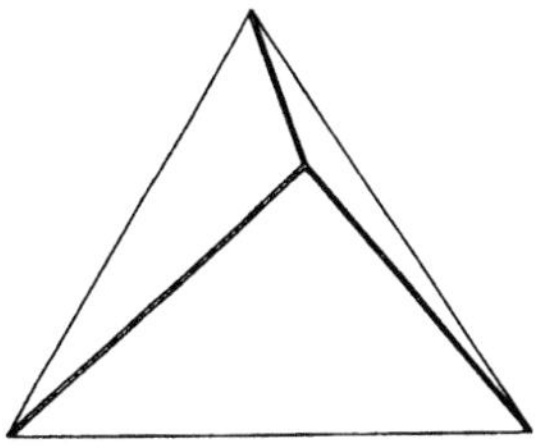

Quasitruncated Tetrahedron

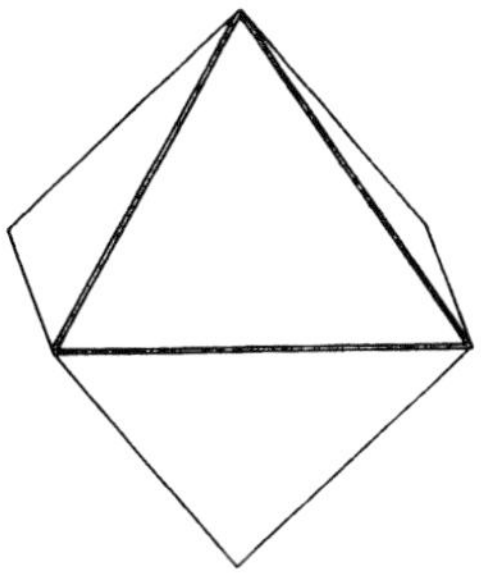

Quasitruncated Octahedron

Quasitruncated Cube

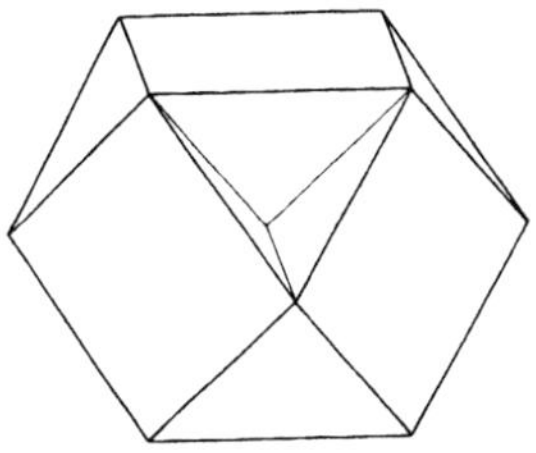

Triquasitruncated Octahedron

Tetraquasitruncated Cuboctahedron

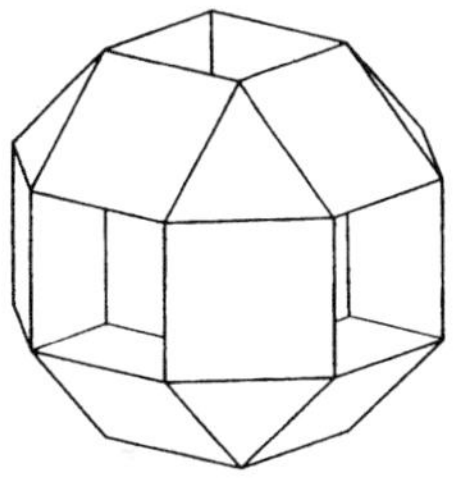

Triquasitruncated Cuboctahedron

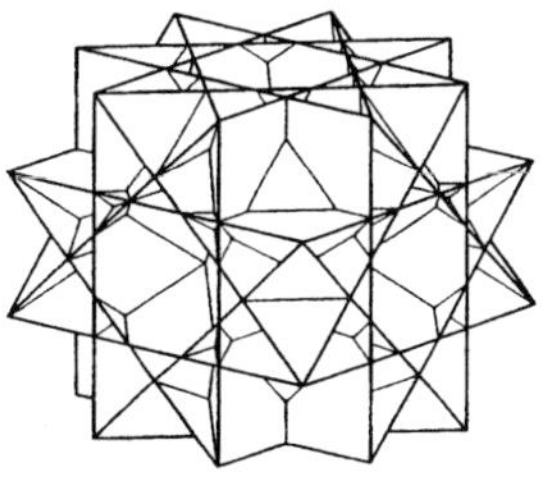

Quasirhombi-Cuboctahedron

Quasiquasitruncated Cuboctahedron

Fig.10: Quasitruncated Polyhedra

QUASITRUNCATED CLOSEPACKS

Looking at the first truncation of each of our two series and substituting quasitruncated solids for truncated ones we find that the resultant assemblies are indeed regular and consistent in themselves despite the effects of interpenetration.

These two quasitruncated closepacks are shown on the right in Figure 12. That for the cubic series is simpler to visualise as each vertex of a truncating octahedron forms around the meeting point of four quasi-truncated cube vertices.

That for the octet series is complicated by the fact that the quasitruncations are effectively the same as the initial solids. We seem to get the original octet closepack again, however careful examination reveals the presence of the truncating cuboctahedra.

Solids	E/V	F/V	S/V
qtruncated cube	5	4 ✱	4
octahedron		4 △	1
$t'\{4,3,4\}$			
Quasitruncated Cubic Closepack			
qtruncated octahedron	12	12 □	6
qtruncated tetrahedron		24 △	8
cuboctahedron			12
$t'\{octet\}$			
Quasitruncated Octetrahedral Closepack			
quasicuboctahedron	8	4 □	4
octahedron		8 △	2
$t'^2\{4,3,4\} = t'^2\{octet\}$			
Quasicuboctahedral			

Table 7: Quasitruncated Closepacks

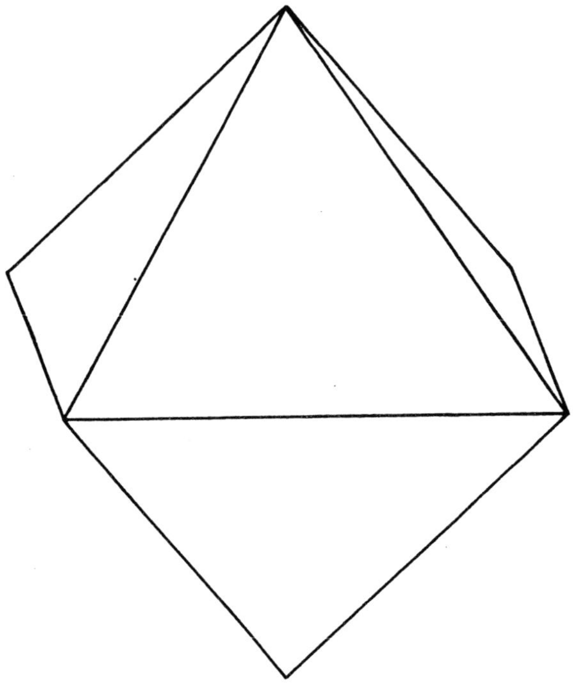

$t'^2\{4,3,4\} = t'^2\{octet\}$
Quasicuboctahedral

Fig.11: Second Quasitruncation

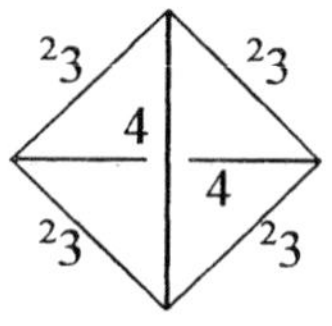

'Quasi-Regular'

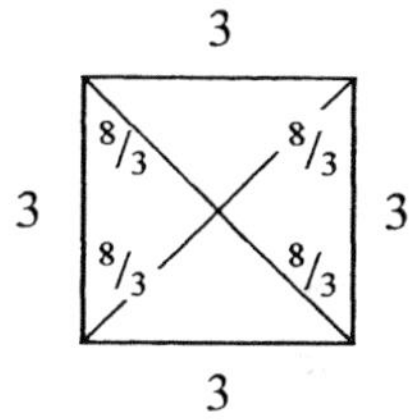

t'{4,3,4}
Quasitruncated Cubic Closepack

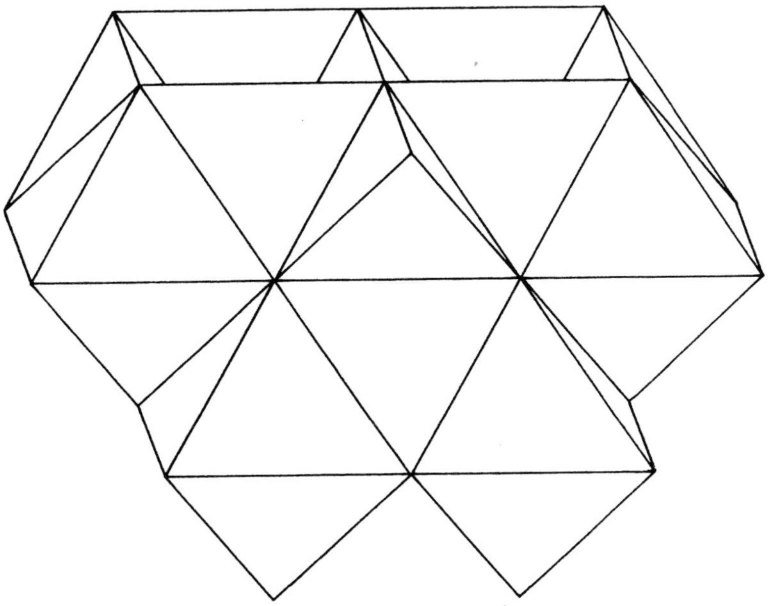

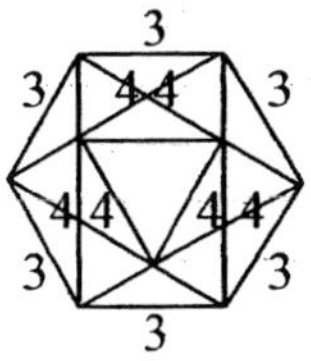

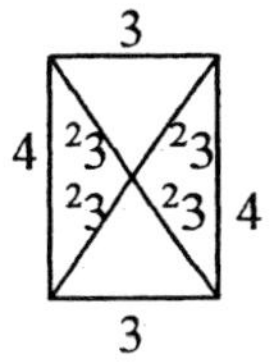

t'{octet}
Quasitruncated Octetrahedral Closepack

Fig.12: Quasitruncated Closepacks First Series

EVEN FACED QUASITRUNCATIONS

Moving on now to the second truncation of each first series, we find as before that {4,3,4} and {octet} yield the same result. The quasicuboctahedral, shown overleaf in Figure 11, has a flat, rather than cuboid, vertex figure, so that what we obtain appears as a single octahedron, but is actually many coincident polyhedra. This situation is the three dimensional analogue of the quasirhombi square tiling that consisted of many squares all piled on top of each other (*Incomplete Tilings* p.24).

A similar situation holds with two of the third quasitruncations of the {4,3,4} series. These are shown opposite in Figure 14 and have vertex figures that are flat rather than tetrahedral and all we obtain appears again as a single polyhedron, that is actually many coincident polyhedra.

Solids	E/V	F/V	S/V
triqtruncated octahedron	4	2 □	2
triqtruncated octahedron		2 ▲	2
$t'^3\{4,3,4\}$		2 ⬡	
Triquasitruncated Octahedral			
qtruncated octahedron	4	2 □	2
qtruncated octahedron		4 ▲	2
$t'^3\{4,3,4\}$			
Quasiquasitruncated Octahedral			
qtruncated octahedron	8	6 □	1
truncated octahedron		4 ▲	4
triqtruncated octahedron		8 ⬡	4
$t'^3\{4,3,4\}$			
Quasitruncated Octahedral Closepack			

Table 8: Quasitruncated Octahedrals

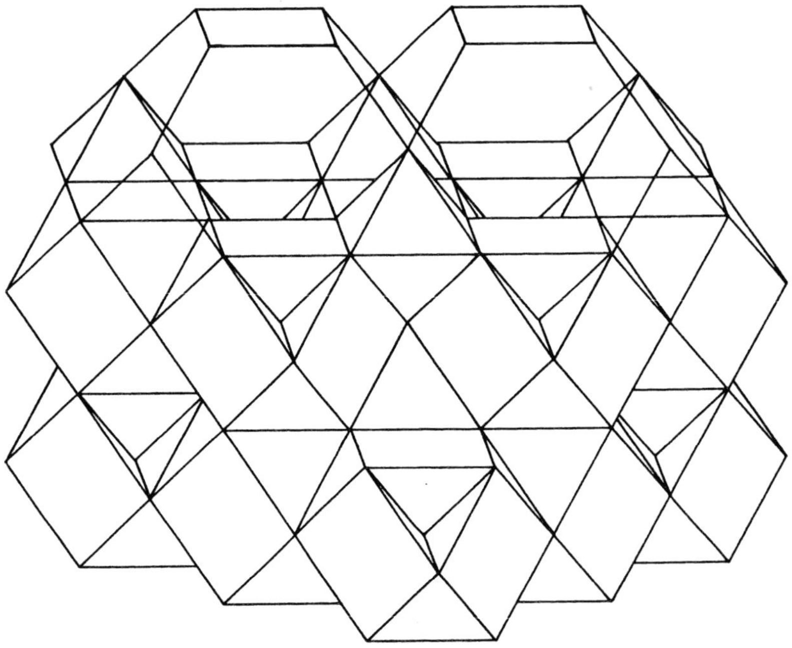

$t'^3\{4,3,4\}$
Quasitruncated Octahedral Closepack

Fig.13: Quasitruncated Closepack

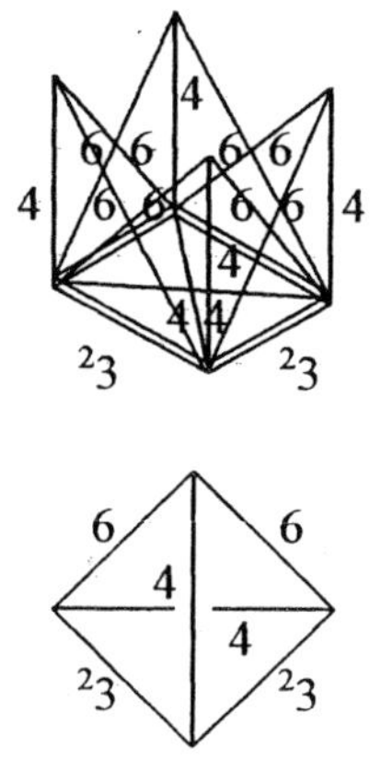

'Even Faced'

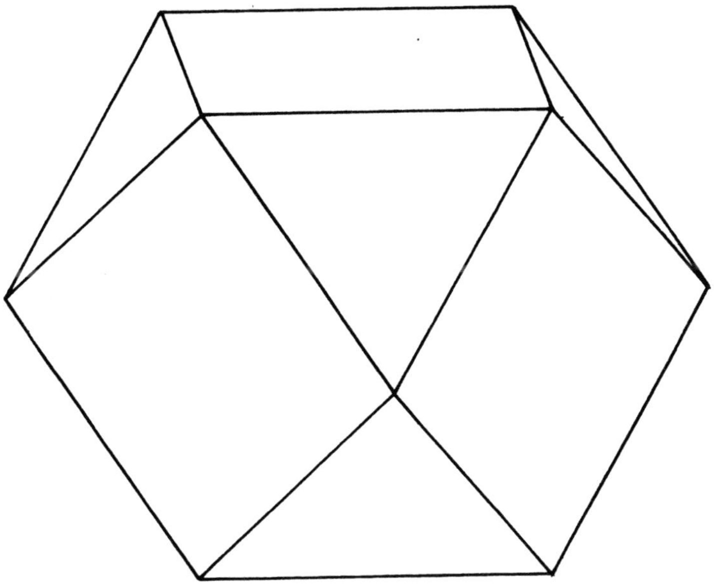

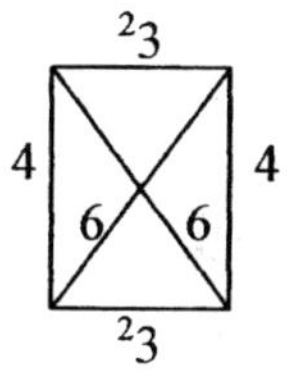

$t'^3\{4,3,4\}$
Triquasitruncated Octahedral

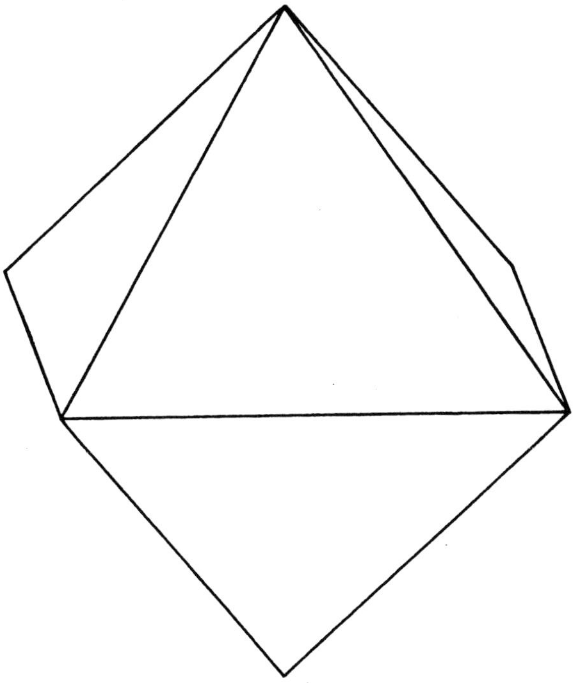

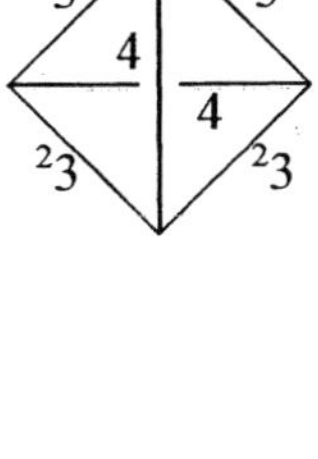

$t'^3\{4,3,4\}$
Quasiquasitruncated Octahedral

Fig.14: Quasitruncated Closepacks 'Even Faced'

MORE EVEN FACED QUASITRUNCATIONS

Back in Figure 13 is shown the only third quasitruncation of {4,3,4} that works, consisting of both quasitruncated and truncated octahedra within a matrix of triquasitruncated octahedra.

Looking now at the other third truncation, that at the centre of the octet series, we find there are again three quasitruncations. The first, quasitruncating just the cube in the original, is shown below in Figure 15.

The other two, one obtained by quasi-truncating just the tetrahedron and the other by quasitruncating both of the original solids, are shown opposite in Figure 16. The second of these involves quasiquasi-truncated cuboctahedra, giving a very dense, much interpenetrated quasi-closepack.

Solids	E/V	F/V	S/V
truncated cube	6	3 □	3
qtruncated tetrahedron		3 △	1
triqtruncated cuboctahedron		6 ⬡	3
t'^{3}{octet}			
Triquasitruncated Cuboctetrahedral Cp			
qtruncated cube	6	3 □	3
qtruncated tetrahedron		3 △	1
qqtruncated cuboctahedron		6 ✲	3
t'^{3}{octet}			
Quasiquasitruncated Cuboctetrahedral Cp			
qtruncated cube	4	1 △	1
truncated tetrahedron		1 □	1
tetraqtruncated cuboctahedron		2 ⬡	2
t'^{3}{octet}		2 ✲	
Tetraquasitruncated Cuboctetrahedral Cp			

Table 9: Quasitruncated Octet Closepacks

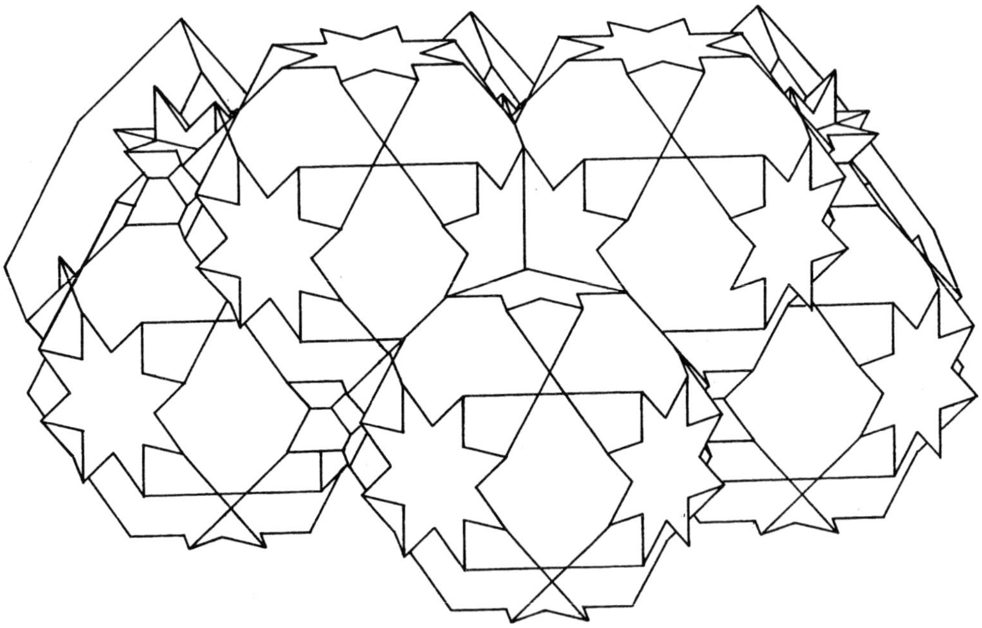

t'^{3}{octet}
Tetraquasitruncated Cuboctetrahedral Closepack

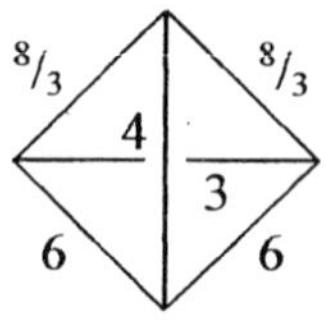

Fig.15: Quasitruncated Closepack 'Even Faced'

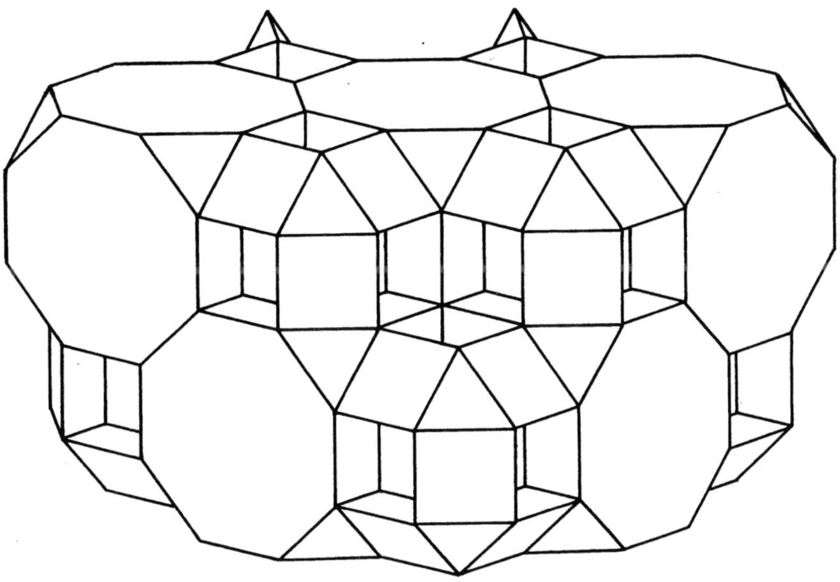

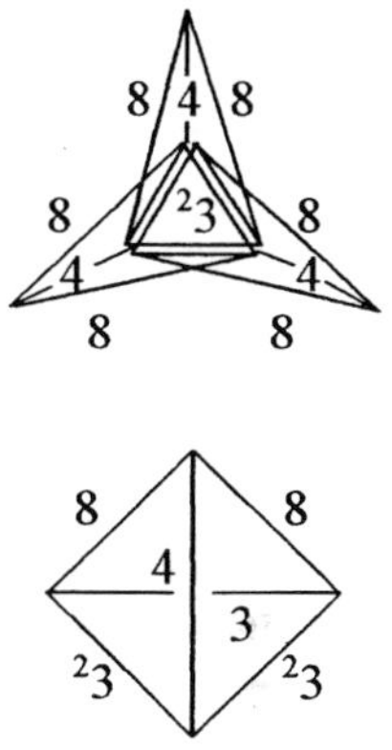

t'^{3} {octet}
Triquasitruncated Cuboctetrahedral Closepack

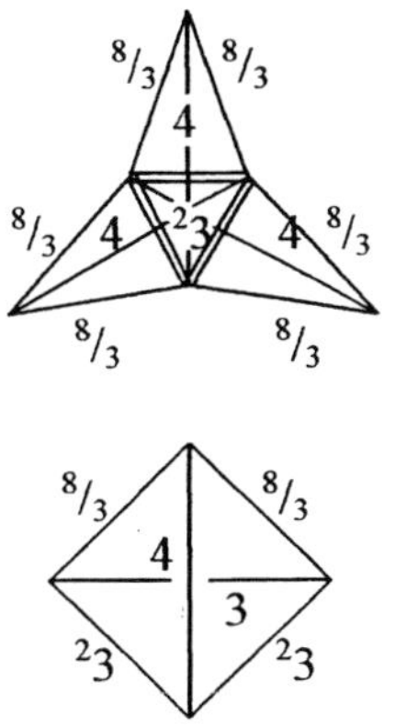

t'^{3} {octet}
Quasiquasitruncated Cuboctetrahedral Closepack

Fig.16: Quasitruncated Closepacks 'Even Faced'

A TRUE QUASIRHOMBI CLOSEPACK

We found earlier when dealing with the true closepacks that the octet series had a fourth truncation (p.12). If we take this as the basis for a new closepack by substituting quasi-rhombihedra into the vertex instead of rhombihedra (as we tried earlier for the cuboctahedral closepack), we can on this occasion succeed to produce more than multiple coincident polyhedra.

The resultant quasirhombi closepack is shown below in Figure 17 and described adjoining at the bottom of Table 10, along with the first two second series quasi-truncations. It is closely allied to the quasiquasitruncated cuboctetrahedral closepack just seen overleaf, although now there are actually no star faces present, just star shaped holes where they would fit. This situation is analogous to that found with some of the quasitruncated tilings.

Solids	E/V	F/V	S/V
truncated cuboctahedron	6	6 □	2
triqtruncated cuboctahedron		1 ◬	1
triqtruncated octahedron		2 ⬡	1
cube		2 ○	2
$t't^2\{4,3,4\} = t't^2\{octet\}$			
Monoquasitruncated Cuboctahedral Cp			
tetraqtr cuboctahedron	6	6 □	2
qqtruncated cuboctahedron		1 ◬	1
triqtruncated octahedron		2 ⬡	1
cube		2 ✵	2
$t't^2\{4,3,4\} = t't^2\{octet\}$			
Diquasitruncated Cuboctahedral Closepack			
cube	6	3 △	1
tetrahedron		6 □	1
quasirhombicuboctahedron			3
$t'^4\{octet\}$			
Quasirhombicuboctetrahedral Closepack			

Table 10: More Quasitruncated Closepacks

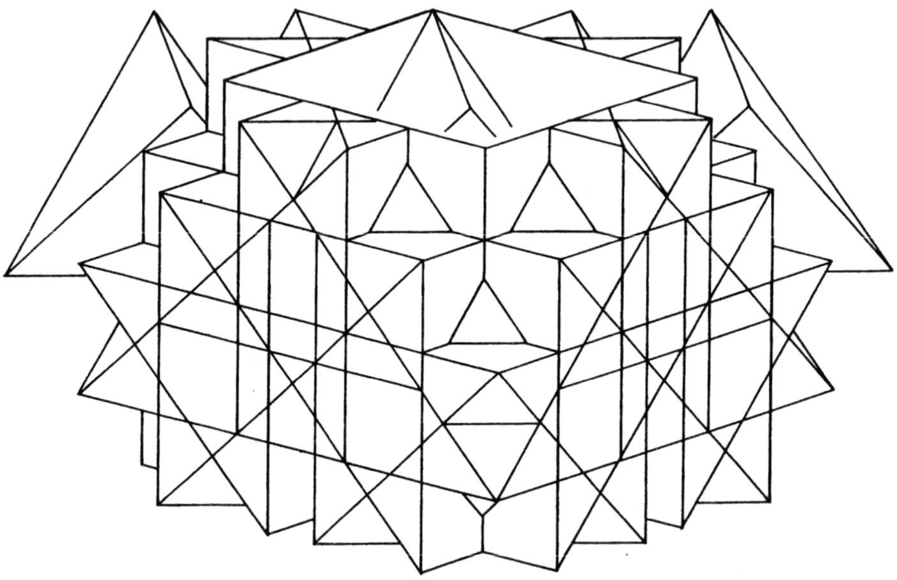

$t'^4\{octet\}$
Quasirhombicuboctetrahedral Closepack

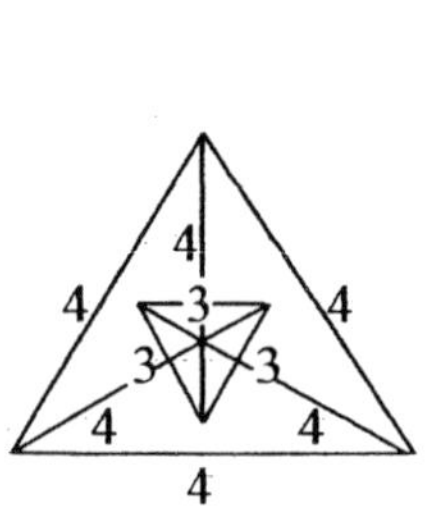

Fig.17: Quasirhombi Closepack

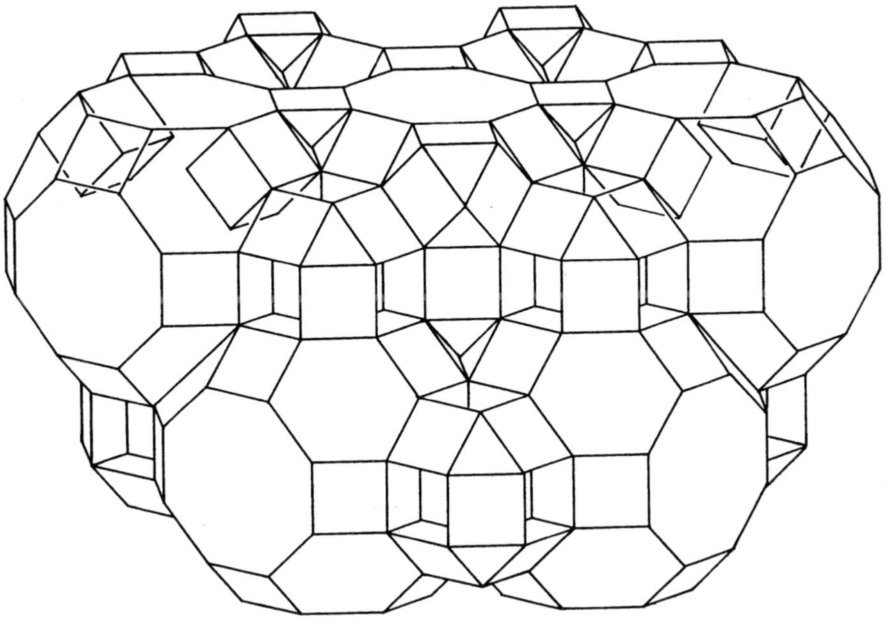

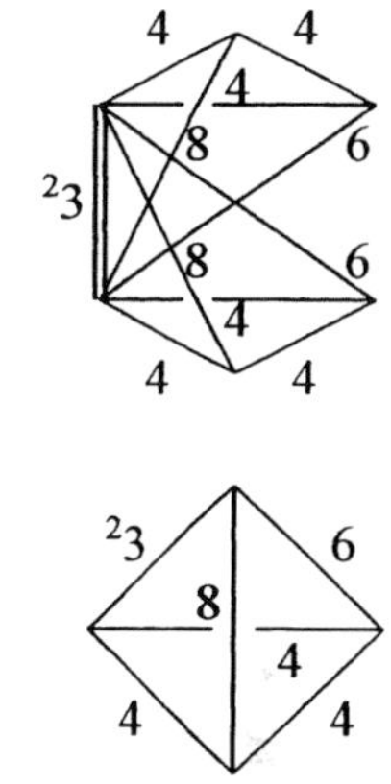

t't^2{4,3,4}
Monoquasitruncated Cuboctahedral Closepack

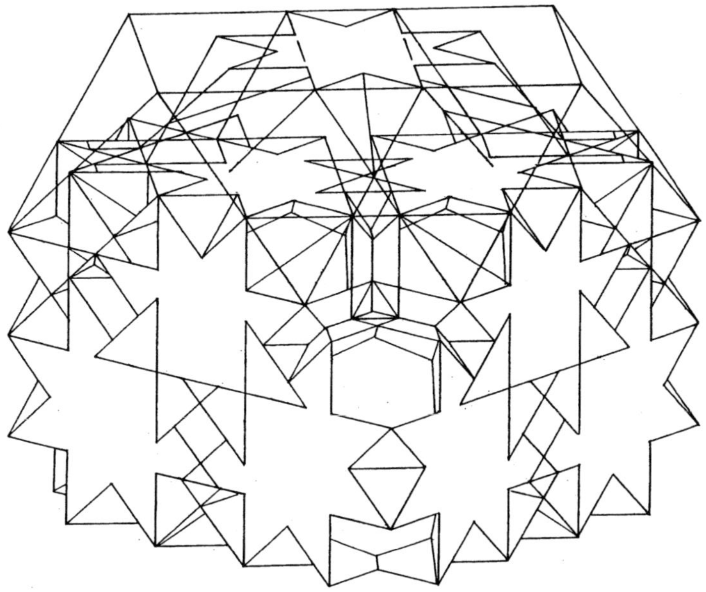

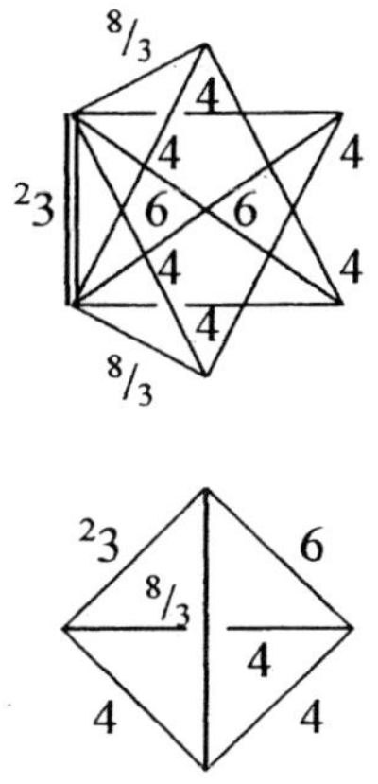

t't^2{4,3,4}
Diquasitruncated Cuboctahedral Closepack

Fig.18: Quasitruncated Closepacks Second series

2nd SERIES QUASITRUNCATIONS

If we now look at the first member of the second series of truncations, the same for both the cubic and octet series, we can find five possible quasitruncated versions of the truncated cuboctahedral closepack seen on page 13.

The two shown overleaf in Figure 18 have a mix of different quasitruncated cubocta-hedra in their make up. These were described in Table 10 facing them, along with the quasirhombicuboctetrahedral closepack illustrated below that table.

The other three quasitruncations of the cuboctahedral closepack, are described in Table 11 adjoining and shown below in Figure 19 and opposite in Figure 20.

Solids	E/V	F/V	S/V
triqtr cuboctahedron	8	10 □	4
qtruncated octahedron		4 △	1
cube		4 ⬡	4
t't^2{4,3,4} = t't^2{octet}			
Triquasitruncated Cuboctahedral Closepack			
qqtr cuboctahedron	8	10 □	4
qtruncated octahedron		4 △	1
cube		4 ✱	4
t't^2{4,3,4} = t't^2{octet}			
Quasiquasitruncated Cuboctahedral Cp			
tetraqtr cuboctahedron	4	3 □	2
truncated octahedron		2 ⬡	1
cube		1 ✱	1
t't^2{4,3,4} = t't^2{octet}			
Tetraquasitruncated Cuboctahedral Cp			

Table 11: Even More Quasitruncations

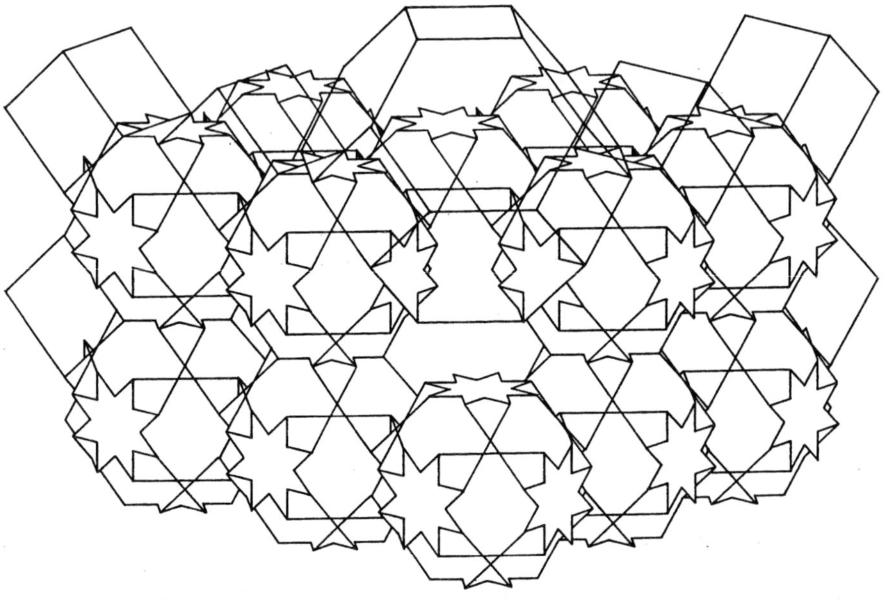

t't^2{4,3,4}
Tetraquasitruncated Cuboctahedral Closepack

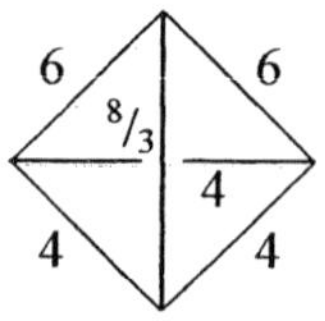

Fig.19: Quasitruncated Closepack

Second series

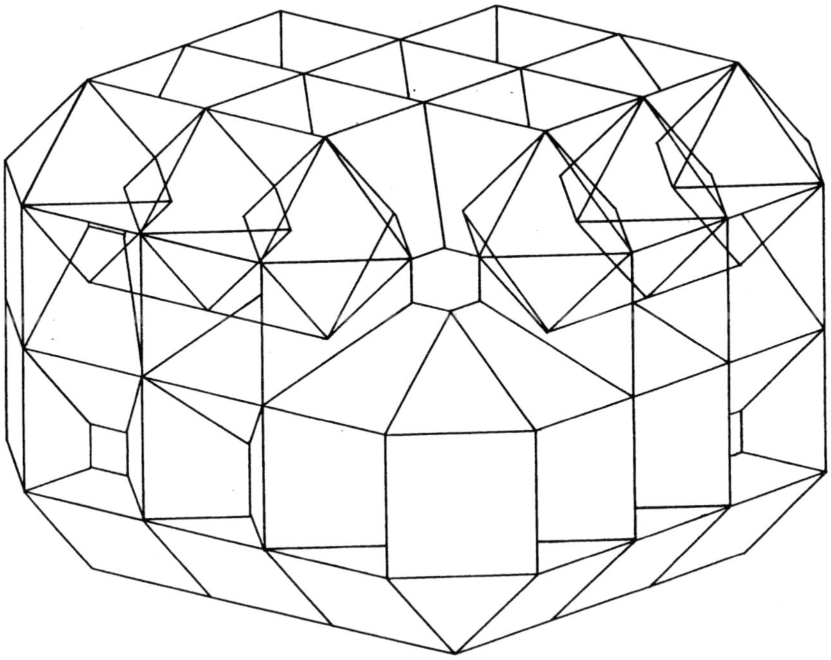

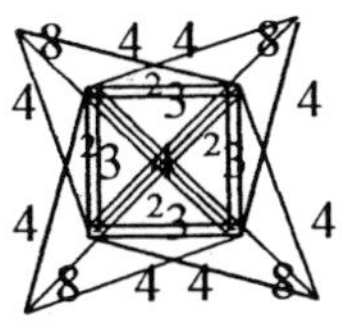

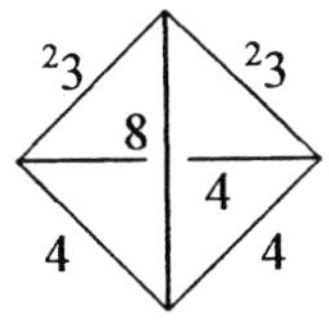

t’t^{2}{4,3,4}
Triquasitruncated Cuboctahedral Closepack

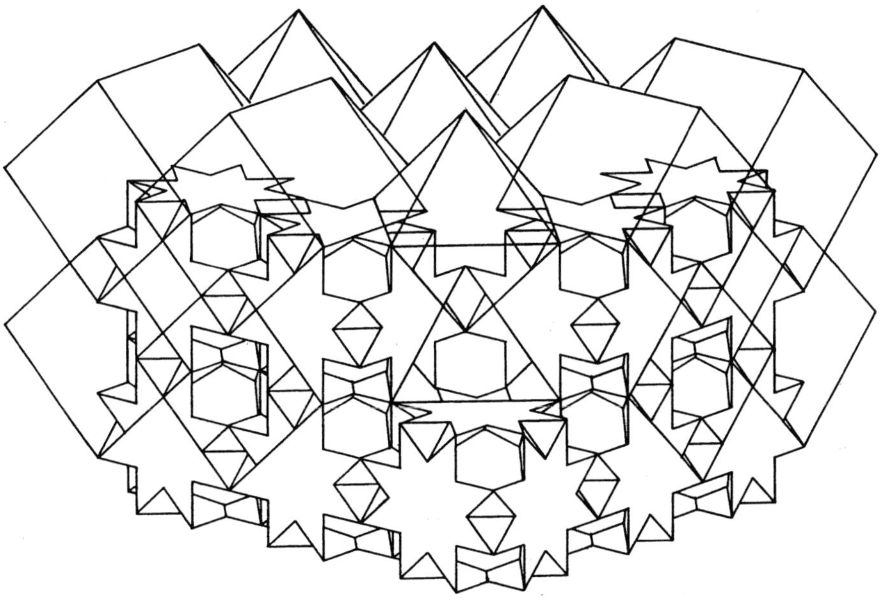

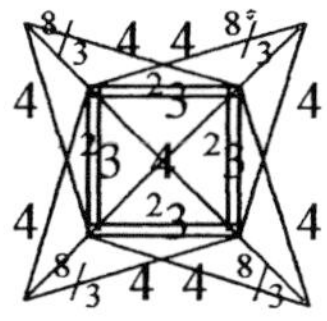

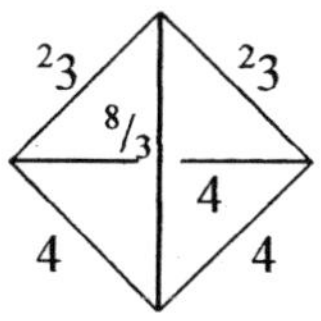

t’t^{2}{4,3,4}
Quasiquasitruncated Cuboctahedral Closepack

Fig.20: Quasitruncated Closepacks Second series

ANOTHER QUASIRHOMBI CLOSEPACK

The second truncation of the second series is again a 'rhombi' form, this time the rhombicuboctahedral closepack also shown on page 13. This too can be replaced with a 'quasirhombi' form by substituting quasirhombihedra within its vertex figure.

The resultant quasirhombicuboctahedral closepack is shown below in Figure 21 and described adjoining at the bottom Table 12, along with the first two quasitruncated odd closepacks, which will be described next.

As with the previous quasirhombi form, this also appears to have star faces, although these are again just the holes in which the octagram stars could fit.

Solids	E/V	F/V	S/V
tetraqtr cuboctahedron	4	3 □	1
truncated cuboctahedron		1 ⬡	1
octaprism		1 ○	1
octagram prism		1 ✵	1
Monoquasitruncated Cuboctaprism Cp			
qqtr cuboctahedron	6	6 □	1
triqtruncated cuboctahedron		1 △	1
octaprism		2 ○	2
octagram prism		2 ✵	2
Diquasitruncated Cuboctaprism Closepack			
cube	6	2 △	2
cuboctahedron		7 □	1
quasirhombicuboctahedron			2
$t'^2t^2\{4,3,4\}$			
Quasirhombicuboctahedral Closepack			

Table 12: Odds and Quasirhombi Closepack

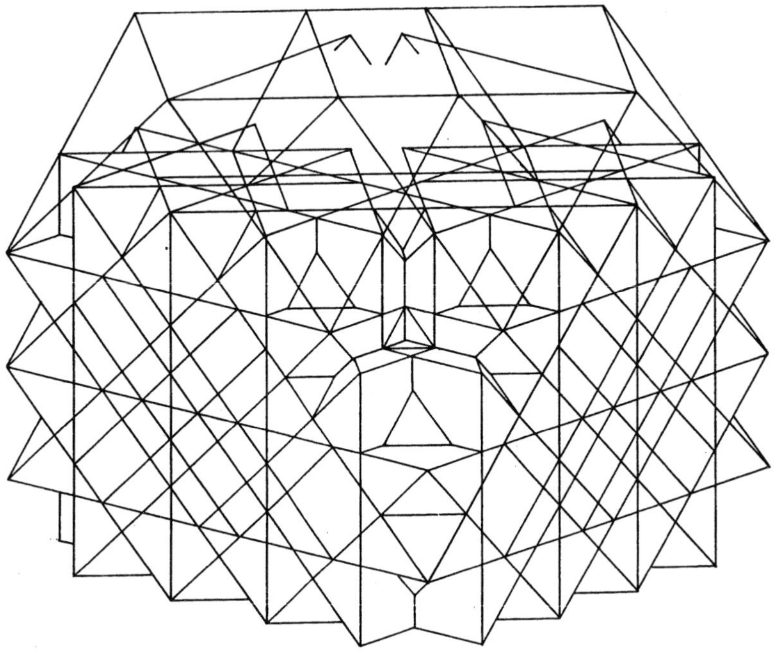

$t'^2t^2\{4,3,4\}$
Quasirhombicuboctahedral Closepack

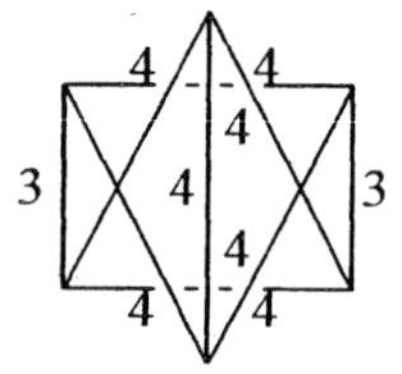

Fig.21: Another Quasirhombi Closepack

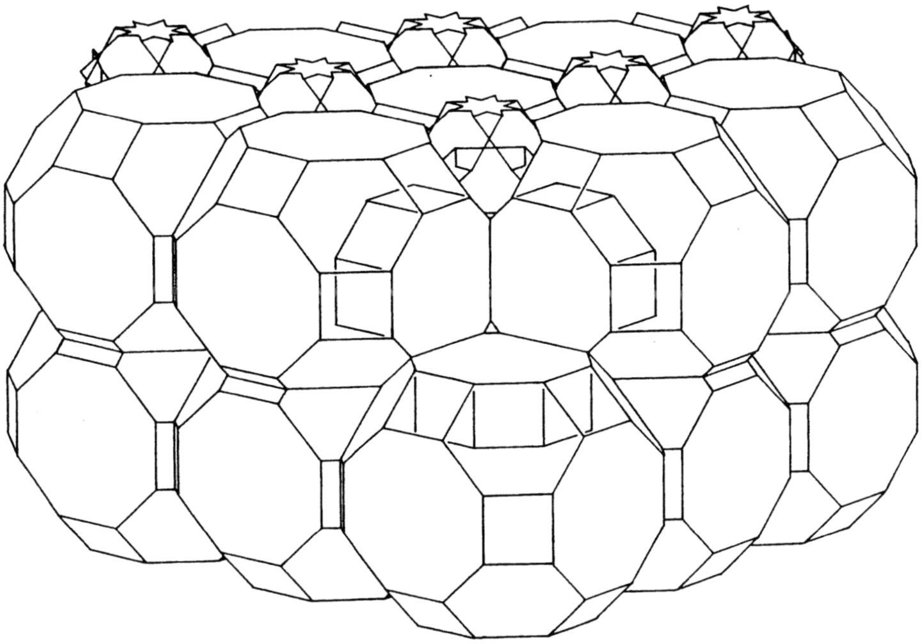

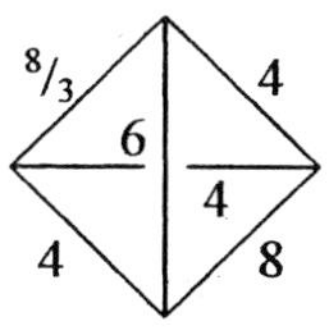

Monoquasitruncated Cuboctaprism Closepack

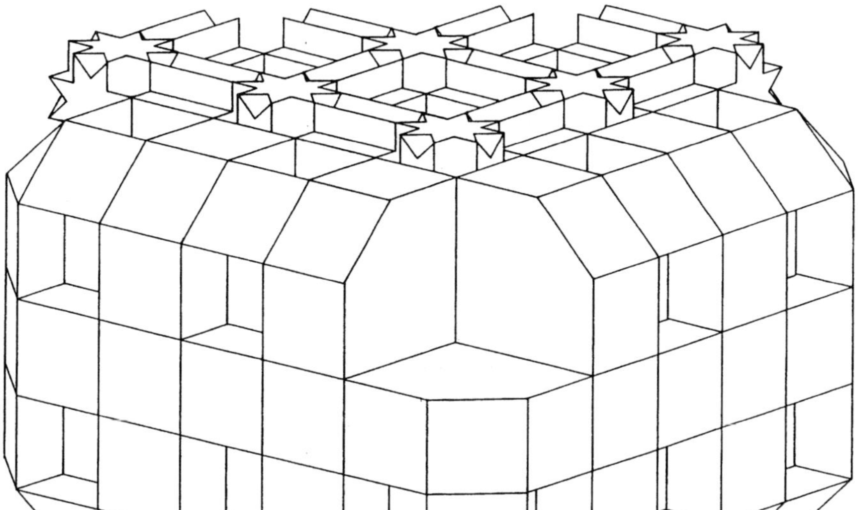

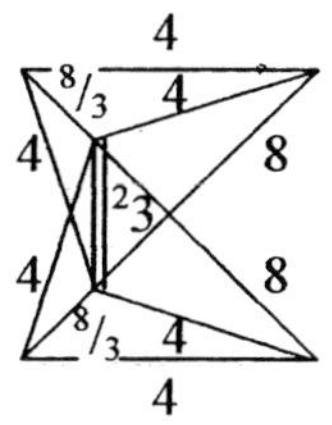

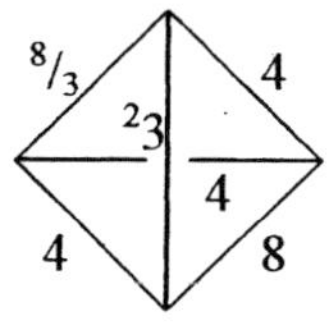

Diquasitruncated Cuboctaprism Closepack

Fig.22: Quasitruncated Odd Closepacks

ODD QUASITRUNCATIONS

If we now examine the odd closepacks with a view to finding quasitruncations, we might at first be discouraged. Replacing truncated forms with quasitruncated in the truncated tetrahedral closepack is rather like our first attempt at a quasirhombi form. The resultant vertex figure is again that of a single polyhedron, this time a repeated tetrahedron (not illustrated), with a two dimensional flat triangular vertex figure.

The other two odd closepacks are a little more forgiving with their prismatic forms. The first with its tetrahedral vertex figure, not unlike a truncated form, gives rise to five further possibilities. Two of these were illustrated overleaf, another is shown below in Figure 23, and the remaining two are degenerate forms with flat vertex figures containing prisms, which appear opposite in Figure 24.

Solids	E/V	F/V	S/V
	4	3 □	
triqtruncated cuboctahedron		1 △	2
octaprism		2 ○	2
Triquasitruncated Cuboctaprism			
	4	3 □	
qqtruncated cuboctahedron		1 △	2
octagram prism		2 ✴	2
Quasiquasitruncated Cuboctaprism			
	4	3 □	
tetraqtruncated cuboctahedron		1 ○	2
octagram prism		2 ✴	2
Tetraquasitruncated Cuboctaprism Cp			

Table 13: Quasitruncated Odd Closepacks

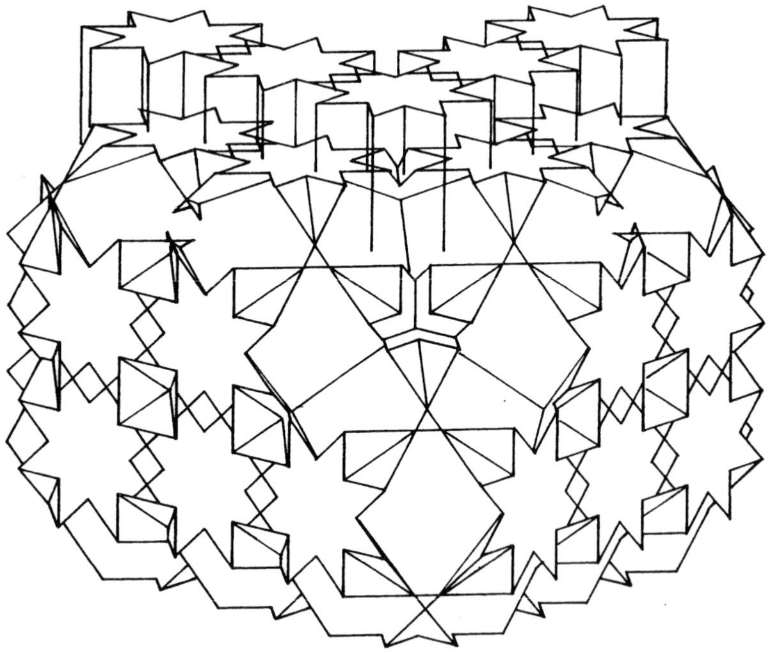

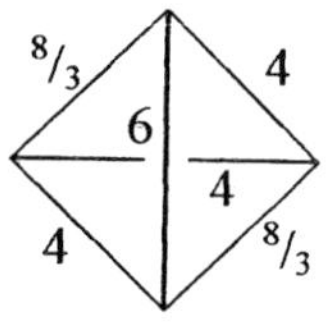

Tetraquasitruncated Cuboctaprism Closepack

Fig.23: Quasitruncated Odd Closepack

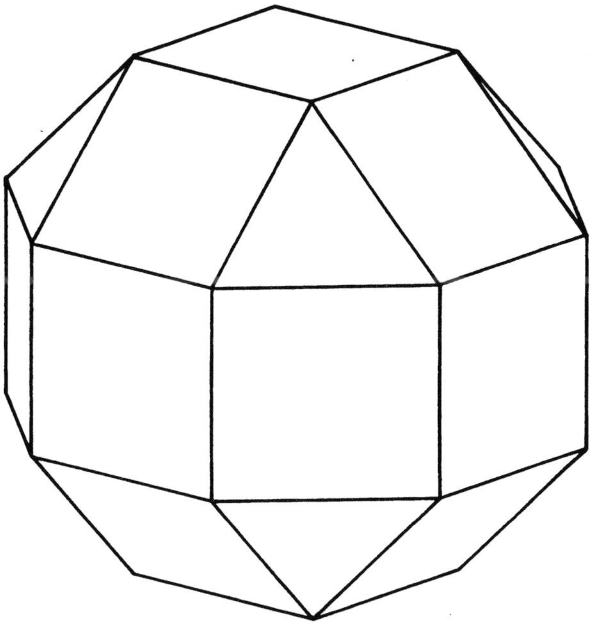

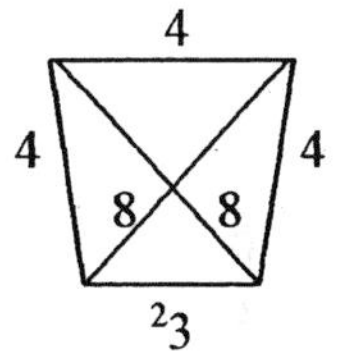

Triquasitruncated Cuboctaprism

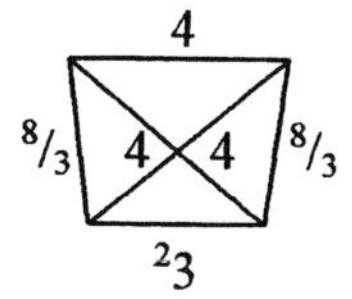

Quasiquasitruncated Cuboctaprism

Fig.24: Quasitruncated Odd Closepacks

QUASIRHOMBIPRISM CLOSEPACKS

The other prismatic odd closepack, the rhombicuboctaprism, contained both truncated cubes and rhombicuboctahedra, each or both of which can be replaced with quasi forms.

There are thus three further quasi-closepacks to consider, which are shown below and on the right in Figures 25 and 26, and described adjoining in Table 14.

The only true star faces in these are those belonging to the quasitruncated cubes and octagram prisms of the two below. The closepack opposite again has only star shaped holes where stars would fit.

Having now looked at all possible true and quasi-closepacks, we must now re-examine them all in the light of another property.

Solids	E/V	F/V	S/V
truncated cube	5	1 △	1
quasirhombicuboctahedron		5 □	1
cube		2 ○	1
octaprism			2
Quasirhombicuboctaprism Closepack			
quasitruncated cube	5	1 △	1
quasirhombicuboctahedron		5 □	1
cube		2 ✵	1
octagram prism			2
Quasirhombicuboctagramprism Closepack			
quasitruncated cube	5	1 △	1
rhombicuboctahedron		5 □	1
cube		2 ✵	1
octagram prism			2
Rhombicuboctagramprism Closepack			

Table 14: Quasirhombiprism Closepacks

Rhombicuboctagramprism Closepack

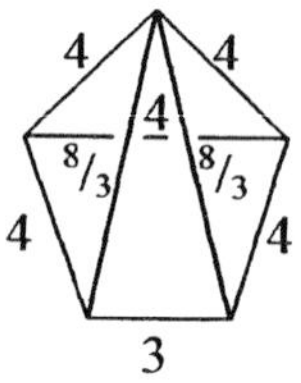

Fig.25: Quasirhombi Odd Closepack

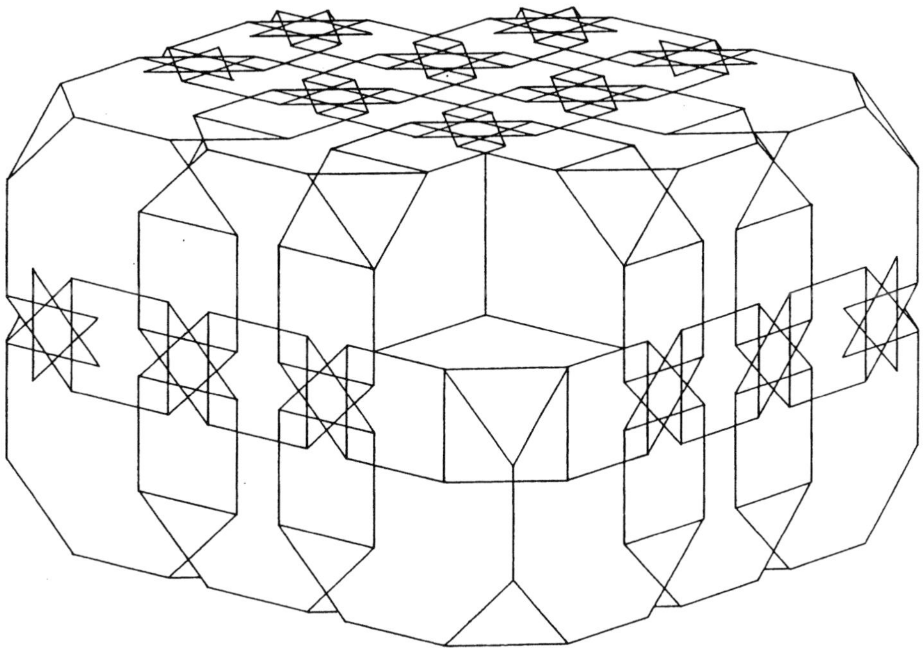

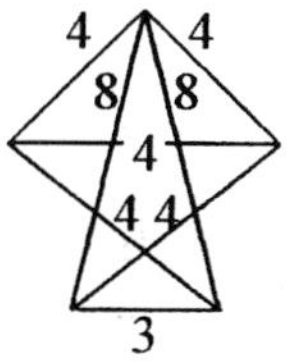

Quasirhombicuboctaprism Closepack

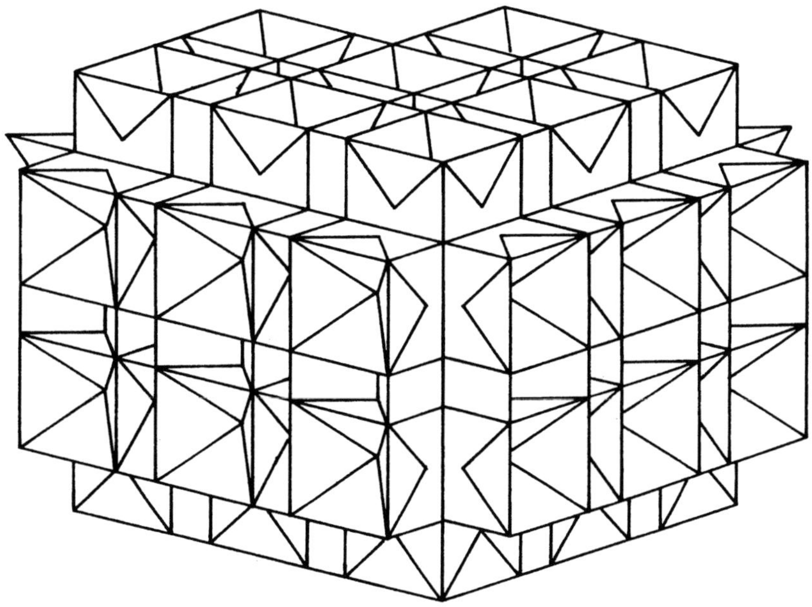

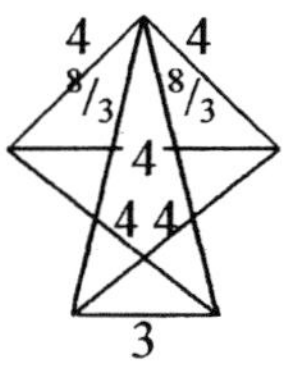

Quasirhombicuboctagramprism Closepack

Fig.26: Quasirhombi Odd Closepacks

CLOSEPACKS WITH TILINGS

We have already seen the simplest closepack that includes a tiling in its make up, that of the cube. Its octahedral vertex figure includes three perpendicular squares, each representing the four-square vertex figure of a square tiling.

The octetrahedral closepack similarly includes the hexagonal vertex figures of four triangle tilings within its vertex figure, whilst within the truncated tetrahedral closepack vertex figure we can find three rectangular vertex figures of the quasi-regular hexagon-triangle tiling.

These three closepacks are shown again to illustrate the presence of these tilings in Figures 27 and 28 adjoining, and described accordingly in Table 15.

Solids	E/V	F/V	S/V
cube	6	12 □	8
square tiling			3
Cubic Closepack			
octahedron	12	24 △	6
tetrahedron			8
triangle tiling			4
Octetrahedral Closepack			
truncated tetrahedron	6	6 △	6
tetrahedron		6 ⬡	2
hexagon-triangle tiling			3
Truncated Tetrahedral Closepack			

Table 15: Closepacks with Tilings

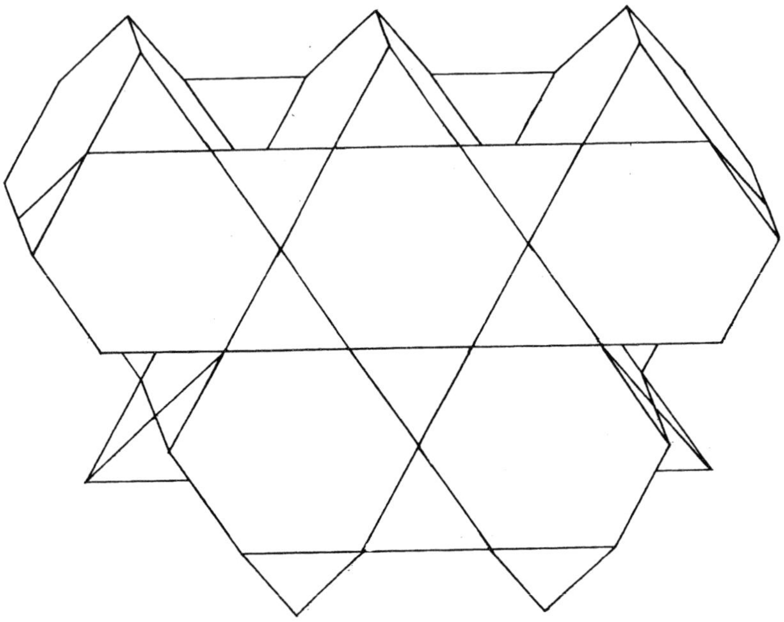

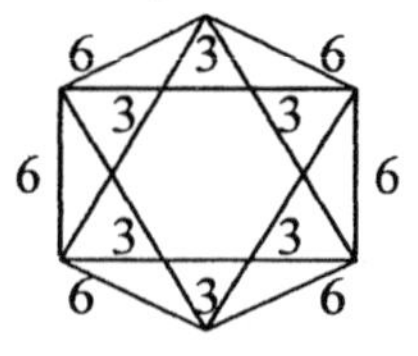

Truncated Tetrahedral Closepack

Fig.27: Closepack with Tilings

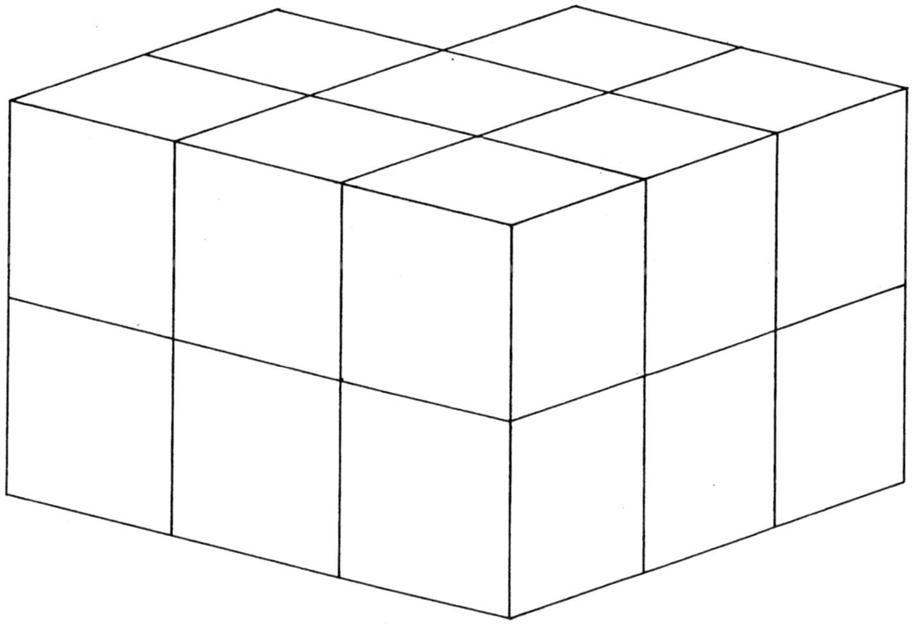

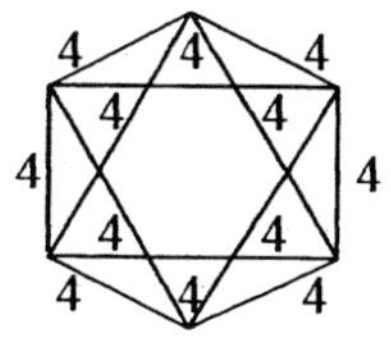

Cubic Closepack

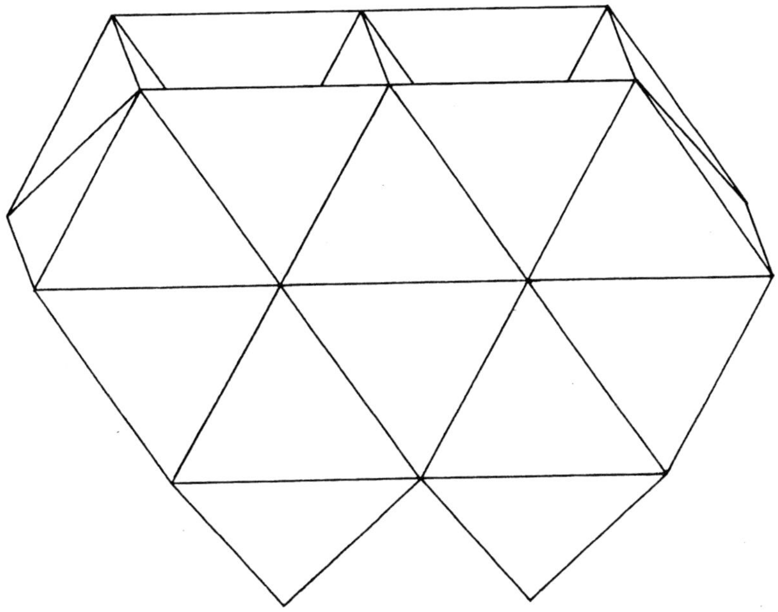

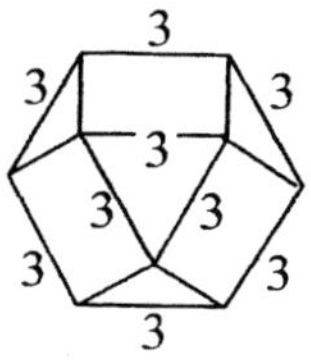

Octetrahedral Closepack

Fig.28: Closepacks with Tilings

HEMITRUNCATED CLOSEPACKS

We can also find tilings within closepacks by subdividing larger vertex figures into smaller ones, most easily done by halving a square pyramidal vertex figure into two tetrahedra, including a base diagonal face.

For instance the inclusion of the internal squares within the octahedra of a truncated cubic closepack creates octagon-square tilings. Similarly the inclusion of internal hexagons within the cuboctahedra of a truncated octetrahedral closepack creates hexagon tilings. The resulting two new closepacks are shown opposite in Figure 30.

Below in Figure 29 is another formed in the same way, this time by taking half of the vertex figure of a rhombicuboctaprism closepack, inserting octagons to create octagon-square tilings.

Solids	E/V	F/V	S/V
truncated cube	5	2 □	4
quasitruncated octahedron		4 ⧋	1
truncated square tiling		4 ⬡	2
Hemitruncated Cubic Closepack			
truncated octahedron	5	2 □	2
truncated tetrahedron		1 ⧋	1
triquasitruncated octahedron		6 ⬡	1
tr triangle (hexagon) tiling			2
Hemitruncated Octetrahedral Closepack			
truncated cube	5	4 □	1
triqtruncated cuboctahedron		1 ⧋	1
octaprism		4 ⬡	2
truncated square tiling			2
Hemirhombicuboctaprism Closepack			

Table 16: Hemitruncated Closepacks

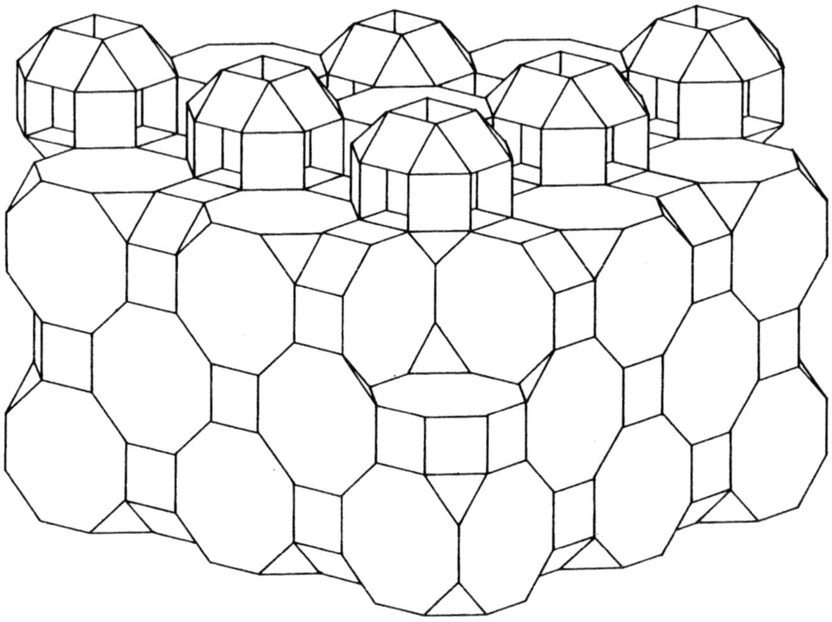

Hemirhombicuboctaprism Closepack

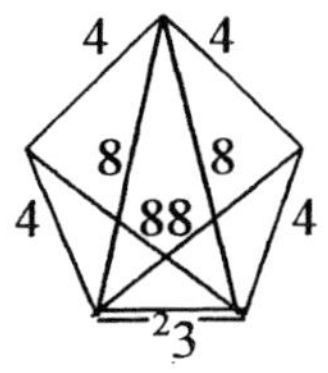

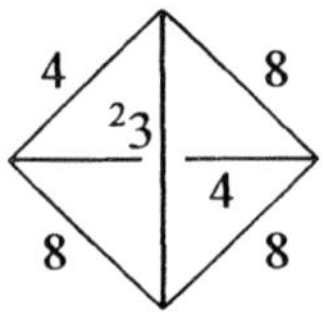

Fig.29: Hemirhombi Odd Closepack

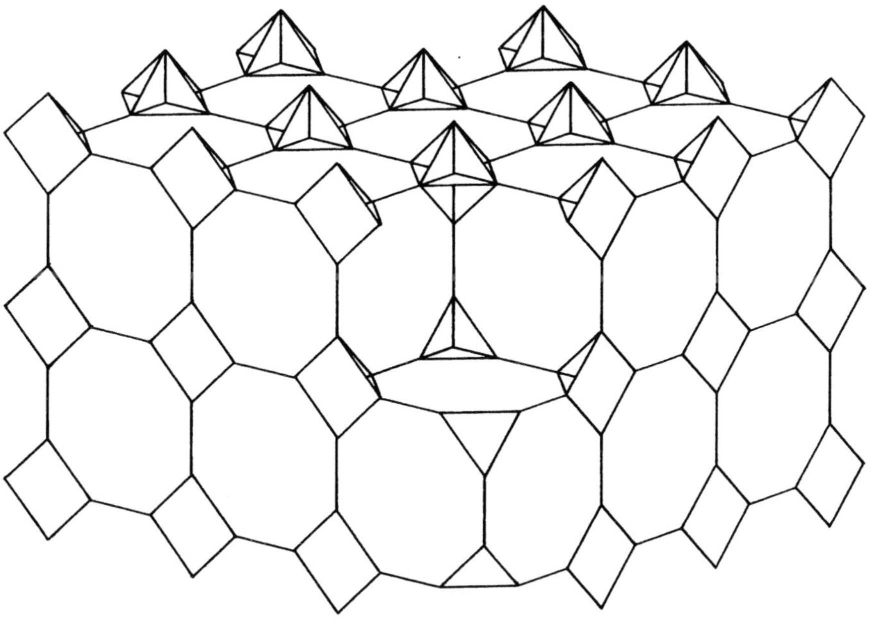

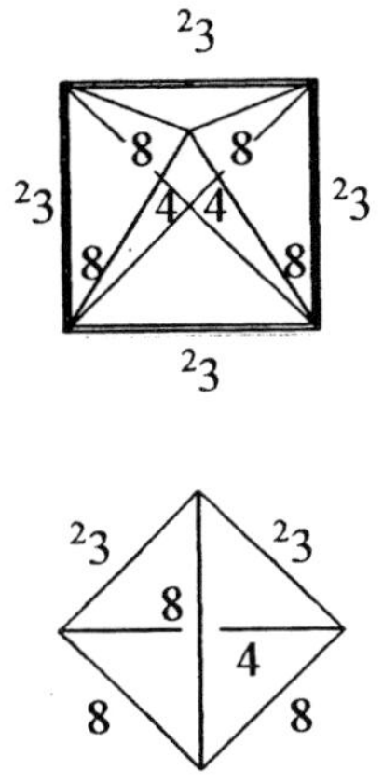

Hemitruncated Cubic Closepack

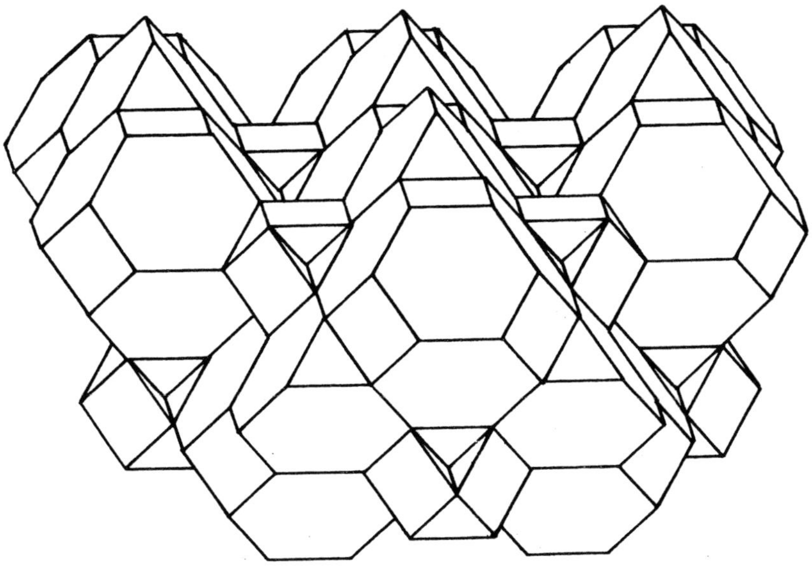

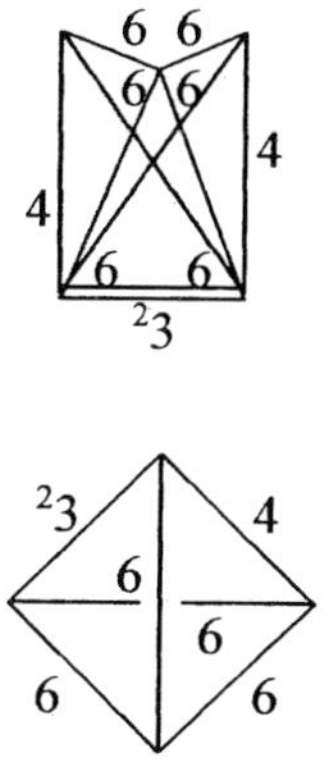

Hemitruncated Octetrahedral Closepack

Fig.30: Hemitruncated Closepacks

Applying this same process to our two quasitruncated closepacks with pyramidal vertex figures, we obtain similar results. These hemiquasitruncated closepacks are shown opposite in Figure 32 and now include within their make up two of the quasitruncated tilings previously seen in *'Incomplete Tilings'* page 21.

Below in Figure 31 is shown the closepack obtained from the other half of the vertex figure of the rhombicuboctaprism close-pack. This inserts octagons by way of a truncated hexahexahedron rather than a triquasitruncated cuboctahedron as was seen overleaf in Figure 29.

This is the first instance we have seen of a {4/2} symmetry amongst the closepacks.

Solids	E/V	F/V	S/V
quasitruncated cube	5	2 □	4
quasitruncated octahedron		4 ◬	1
qqtruncated square tiling		4 ✲	2
Hemiquasitruncated Cubic Closepack			
qtruncated octahedron	12	12 □	6
quasitruncated tetrahedron		24 ◬	8
triquasitruncated octahedron		24 ○	24
quasitruncated triangle tiling			4
Hemiquasitruncated Octetrahedral Cp			
tr hexahexahedron	5	4 □	1
tr stellated hexahedron		1 ⧈	1
octaprism		4 ○	2
truncated square tiling			2
Hemihexahexoctaprism Closepack			

Table 17: Hemiquasitruncated Closepacks

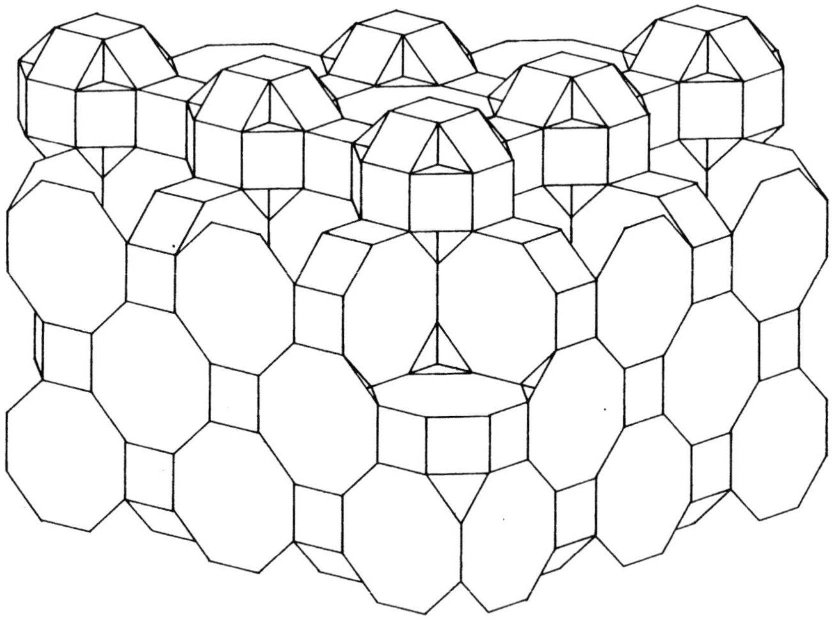

Hemihexahexoctaprism Closepack

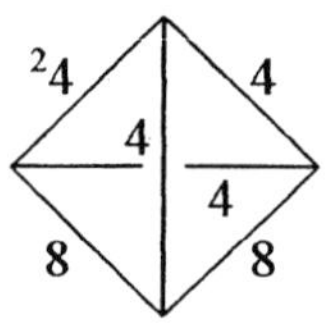

Fig.31: Hexahexaprism Closepack

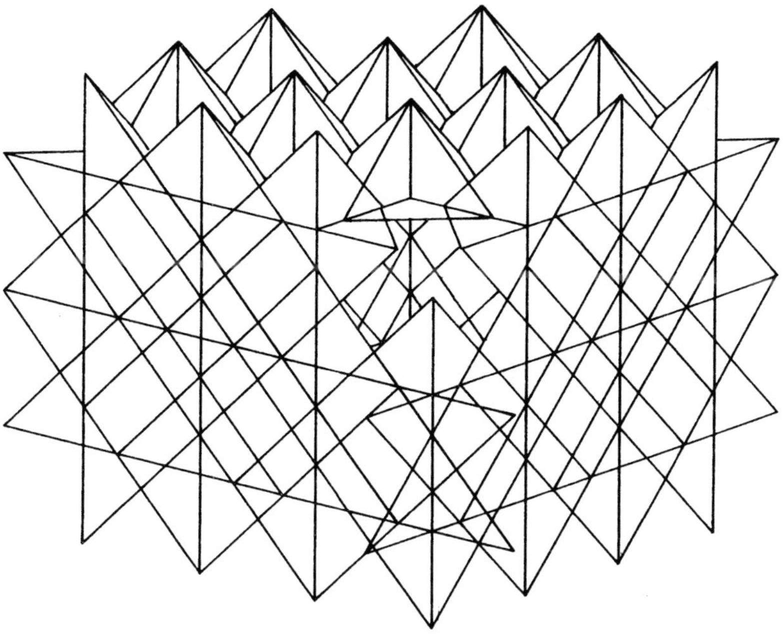

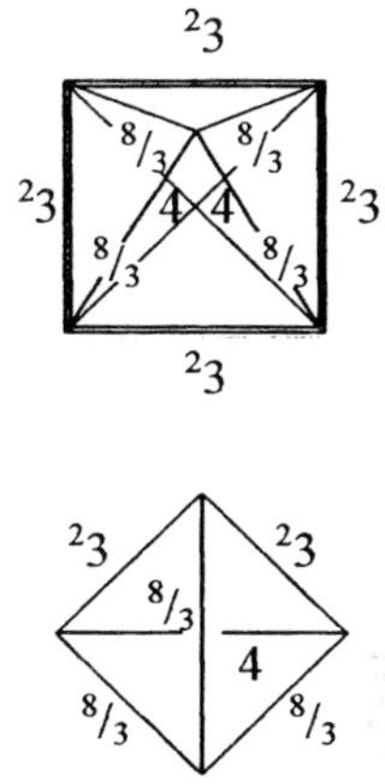

Hemiquasitruncated Cubic Closepack

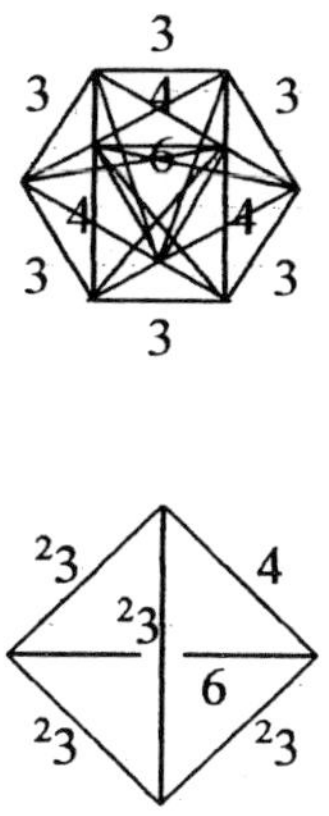

Hemiquasitruncated Octetrahedral Closepack

Fig.32: Hemiquasitruncated Closepacks

HEMIRHOMBIPRISM CLOSEPACKS

The other quasirhombiprism closepacks, described on page 32, also yield further 'hemi' versions by taking half their pyramidal vertex figures as tetrahedral vertex figures in their own right. The results are shown below and opposite in Figures 33 and 34 and described in Table 18 adjoining.

Whilst we have ostensibly been inserting extra faces at each vertex to make the hidden tiling complete, we have also been doing something else since that extra face is invariably one latent within one of the polyhedra in the vertex figure of the original closepack. We have thus in each case also been substituting certain quasi-truncated polyhedra for ordinary ones that share the same shells. These substitutions can however be carried out in two different ways (compare Figures 29 and 31).

Solids	E/V	F/V	S/V
truncated cube	5	4 □	1
qqtruncated cuboctahedron		1 △	1
octaprism		2 ○	2
quasitruncated square tiling		2 ✵	2
Hemiquasirhombicuboctaprism Closepack			
quasitruncated cube	5	4 □	1
qqtruncated cuboctahedron		1 △	1
octagramprism		4 ✵	2
qqtruncated square tiling			2
Hemiquasirhombicuboctagramprism Cp			
quasitruncated cube	5	4 □	1
triqtruncated cuboctahedron		1 △	1
octagramprism		2 ○	2
quasitruncated square tiling		2 ✵	2
Hemirhombicuboctagramprism Closepack			

Table 18: Hemirhombiprism Closepacks

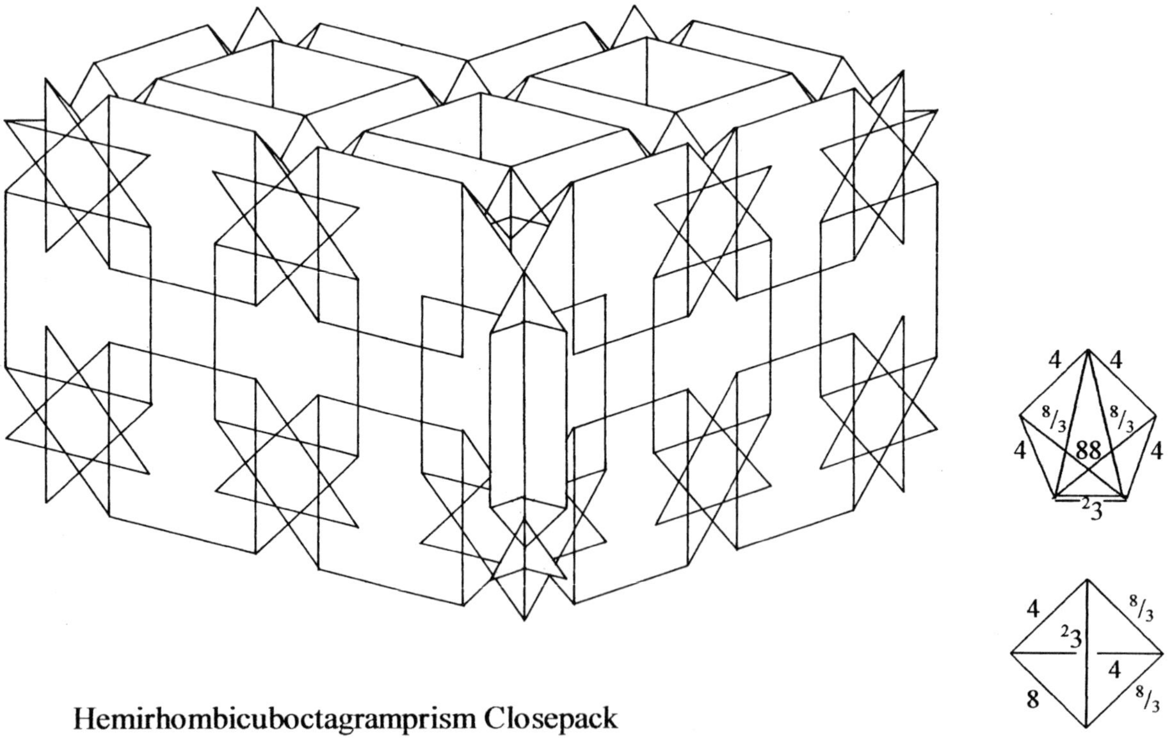

Hemirhombicuboctagramprism Closepack

Fig.33: Hemirhombi Odd Closepack

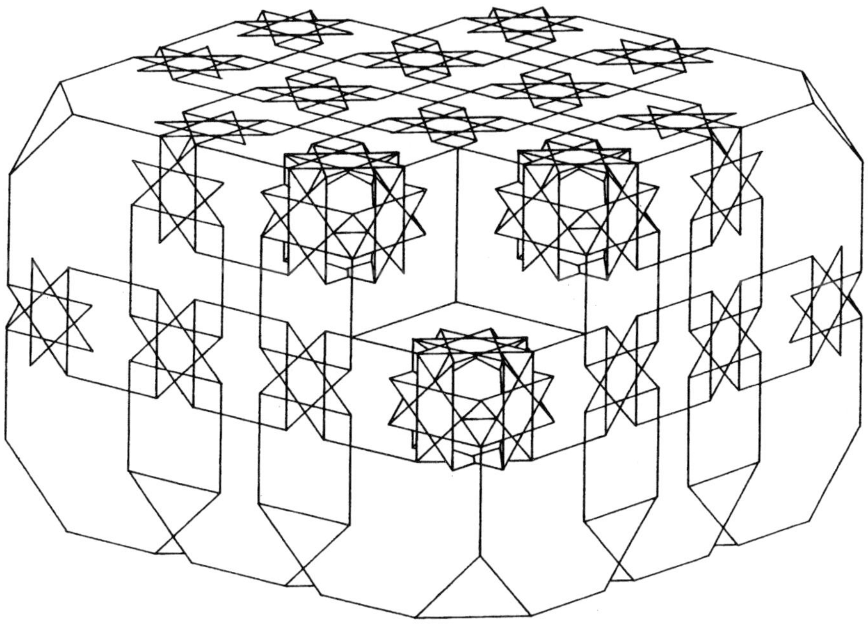

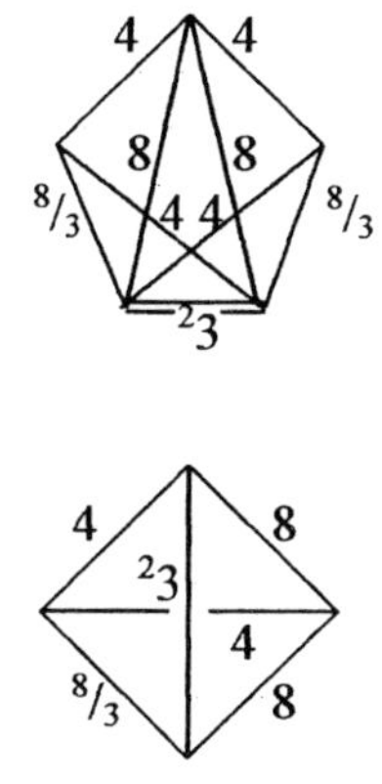

Hemiquasirhombicuboctaprism Closepack

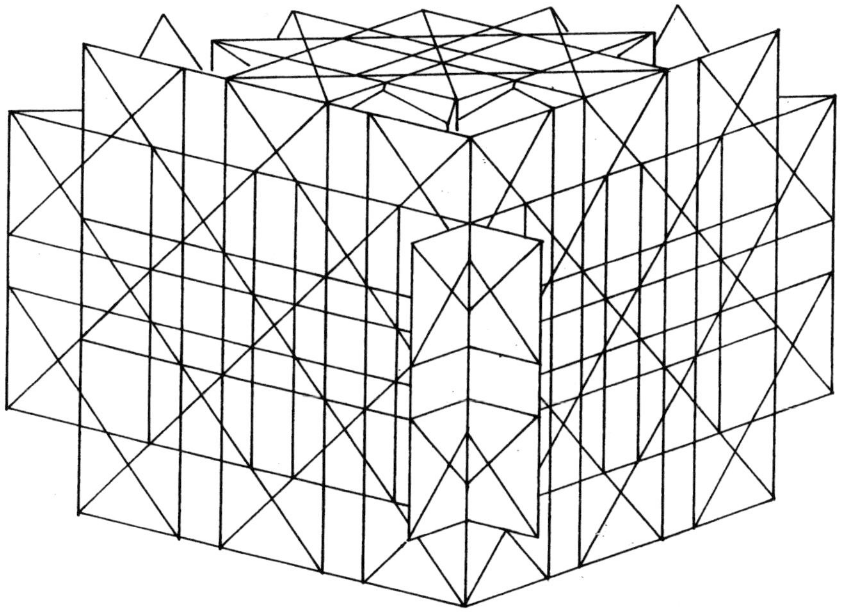

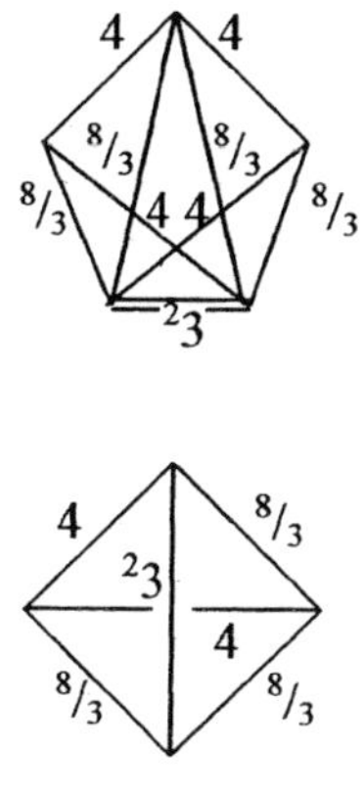

Hemiquasirhombicuboctagramprism Closepack

Fig.34: Hemiquasirhombi Odd Closepacks

HEXAHEXAPRISM CLOSEPACKS

If we now examine the other half of the quasirhombiprism closepack vertex figures we have just dissected we can find a further closepack in each case. Again these include truncated or quasitruncated hexahexahedra with {4/2} symmetries rather than the various truncated cuboctahedra we have previously seen.

The resultant closepacks, each of which contains a tiling, are shown below and opposite in Figures 35 and 36 and described adjoining in Table 19.

Having looked at all the pyramidal vertex figures described thus far, we will next look at some larger vertex figures and see what we can find within.

Solids	E/V	F/V	S/V
quasitr hexahexahedron	5	4 □	1
tr stellated hexahedron		1 ⧈	1
octaprism		2 ⯃	2
quasitruncated square tiling		2 ✵	2
Hemiquasihexahexoctaprism Closepack			
quasitr hexahexahedron	5	4 □	1
tr stellated hexahedron		1 ⧈	1
octagramprism		4 ✵	2
quasiquasitruncated square tiling			2
Hemiquasihexahexoctagramprism Cp			
tr hexahexahedron	5	4 □	1
tr stellated hexahedron		1 ⧈	1
octagramprism		2 ⯃	2
quasitruncated square tiling		2 ✵	2
Hemihexahexoctagramprism Closepack			

Table 19: Hexahexaprism Closepacks

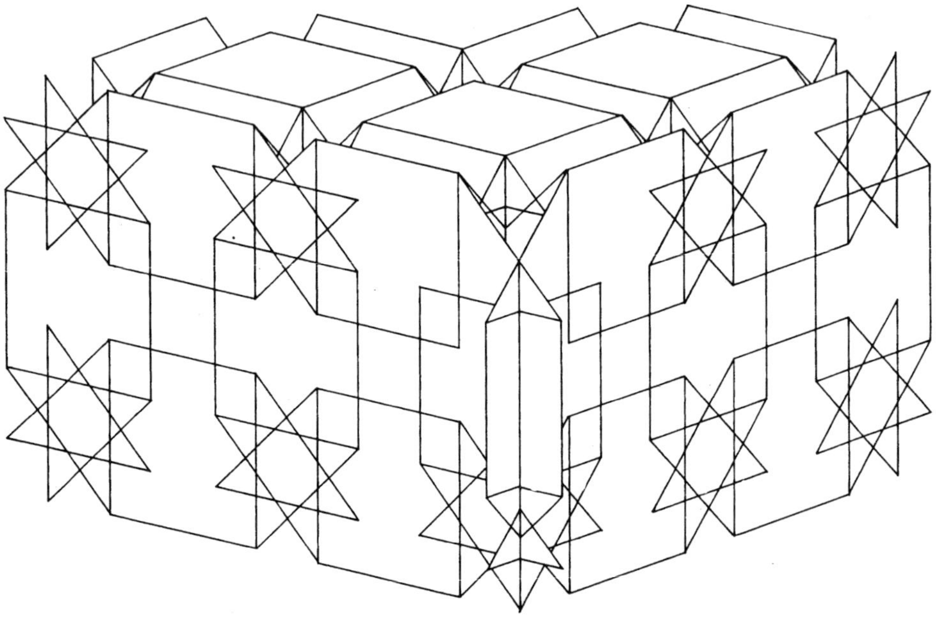

Hemihexahexoctagramprism Closepack

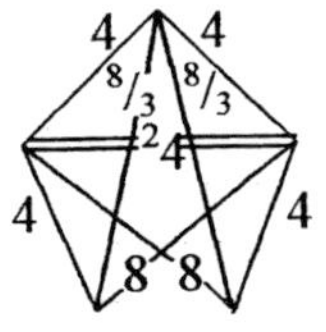

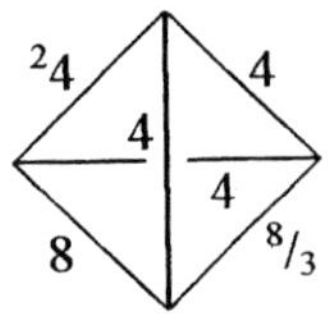

Fig.35: Hexahexaprism Closepack

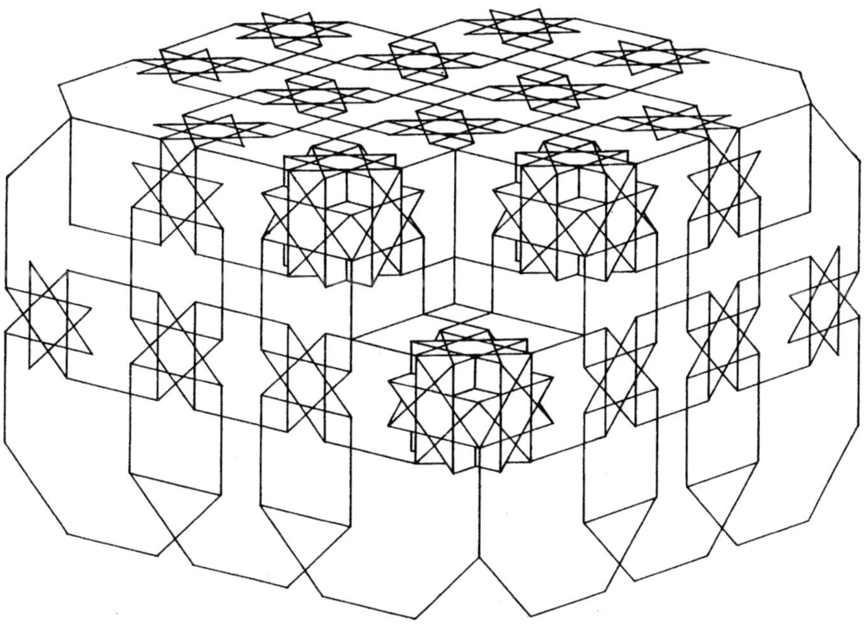

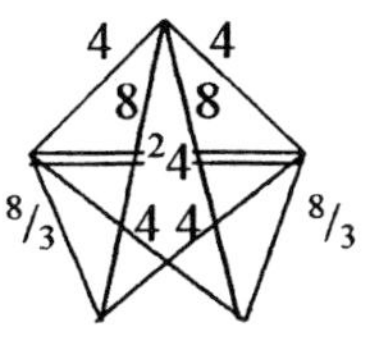

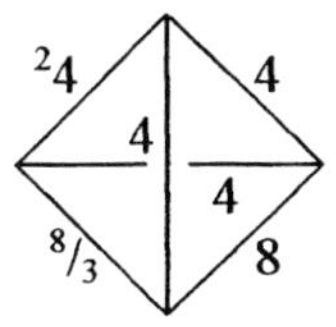

Hemiquasihexahexoctaprism Closepack

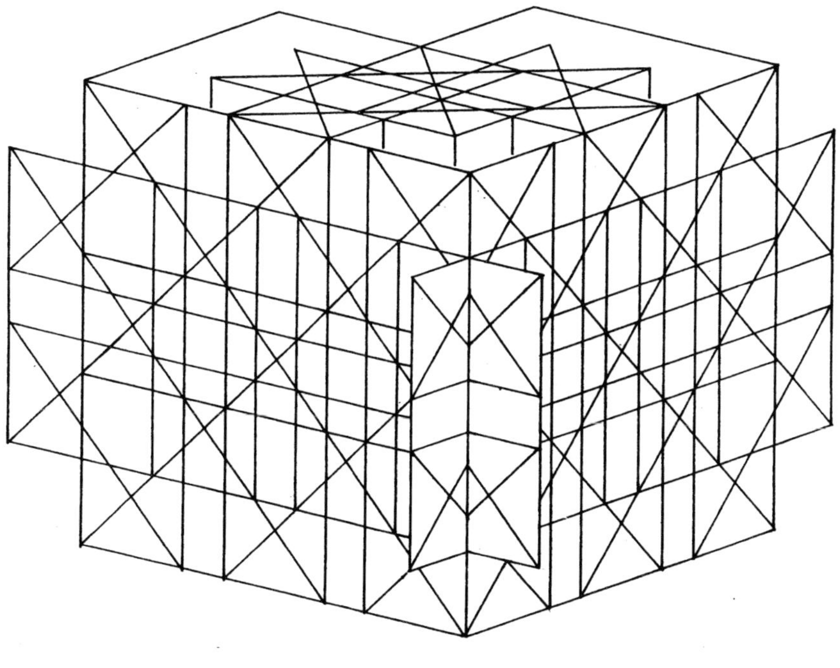

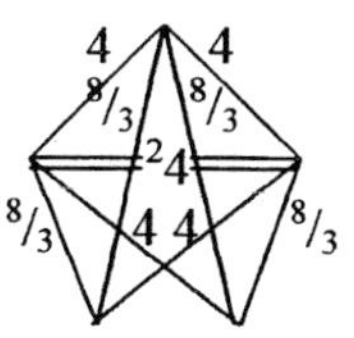

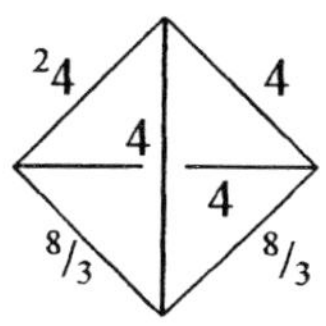

Hemiquasihexahexoctagramprism Closepack

Fig.36: Quasihexahexaprism Closepacks

RHOMBICUBOCTETRAHEDRALS

The next largest size of vertex figure we can look into after the pyramidal ones is the triangular prism type.

Starting with the rhombicuboctetrahedral closepack (p.12), this can be subdivided into one tetrahedral and one pyramidal vertex figure or alternatively at most three tetrahedral vertices.

Of these possibilities, we have already seen the triquasitruncated cuboctetrahedral (p.23) and its other half the rhombicuboctaprism closepack (p.15), the latter breaking down into the hemirhombicuboctaprism (p.36) and the hemihexahexoctaprism (p.38) closepacks just seen.

There is however one with a similar vertex figure to this last, not seen before, which is shown below in Figure 37.

Solids	E/V	F/V	S/V
rhombicuboctahedron	6	6 □	1
truncated cuboctahedron		1 ◬	2
cube		1 ⧈	2
truncated hexahexahedron		2 ⬡	1
triquasitruncated octahedron		2 ○	1
Rhombicuboctahexahedral Closepack			
cuboctahedron	6	6 □	1
triqtruncated cuboctahedron		2 ◬	2
cube		4 ○	2
truncated square tiling			2
Cuboctacuboctahedral Closepack			
tr hexahexahedron	6	3 □	3
tr stellated hexahedron		3 ⧈	1
truncated square tiling		6 ○	3
Hemirhombicuboctetrahedral Closepack			

Table 20: Rhombicuboct(etr)ahedrals

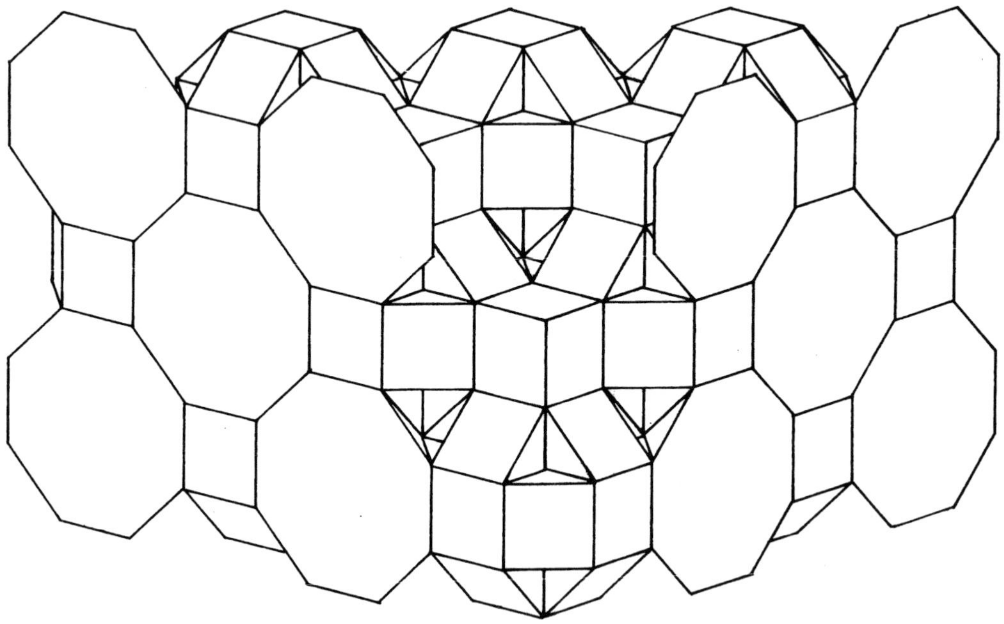

Hemirhombicuboctetrahedral Closepack

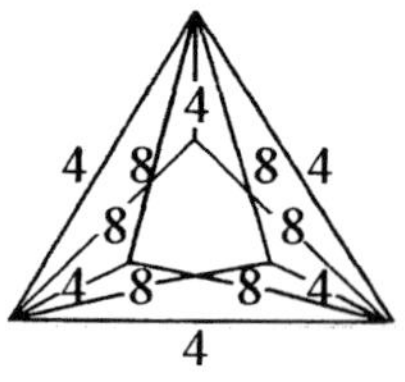

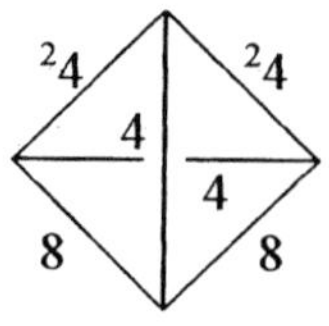

Fig.37: Rhombicuboctetrahedral Closepack

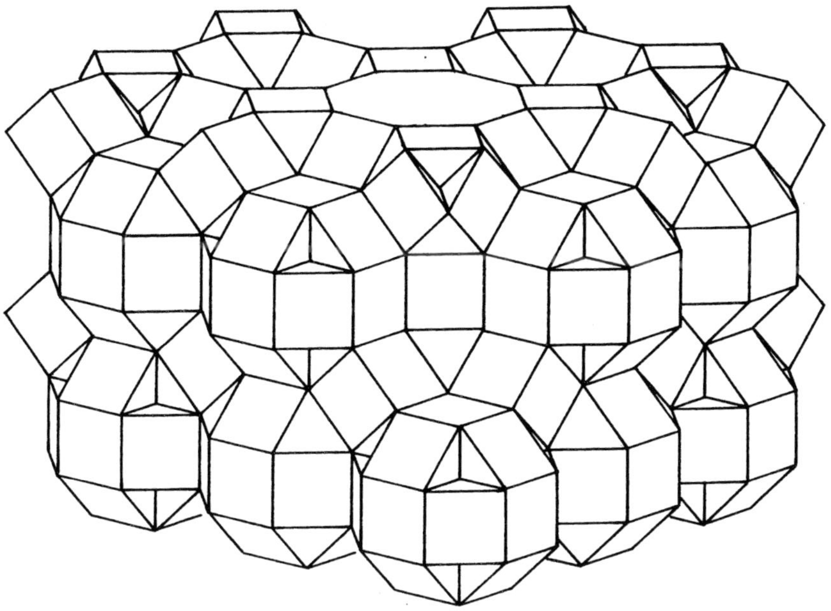

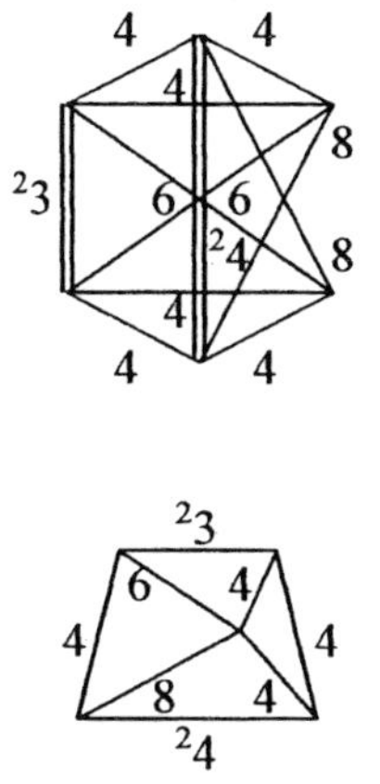

Rhombicuboctahexahedral Closepack

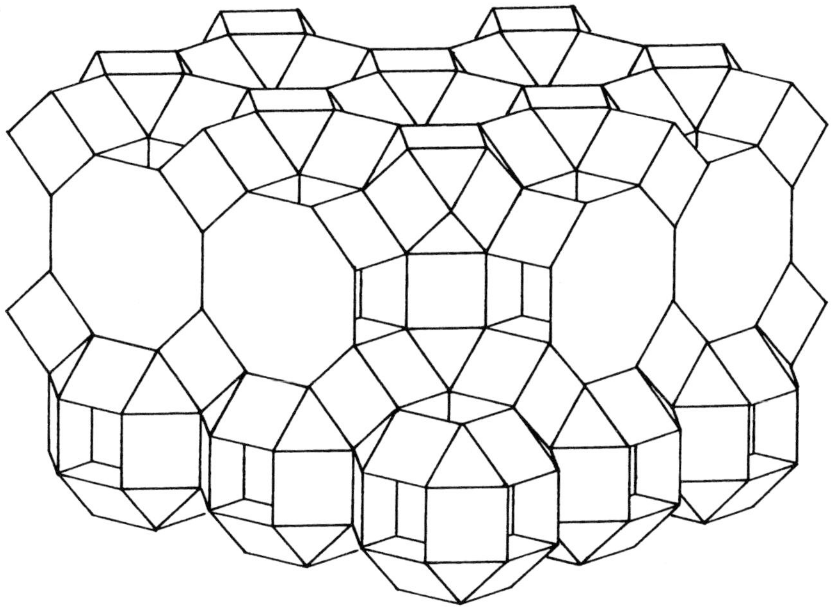

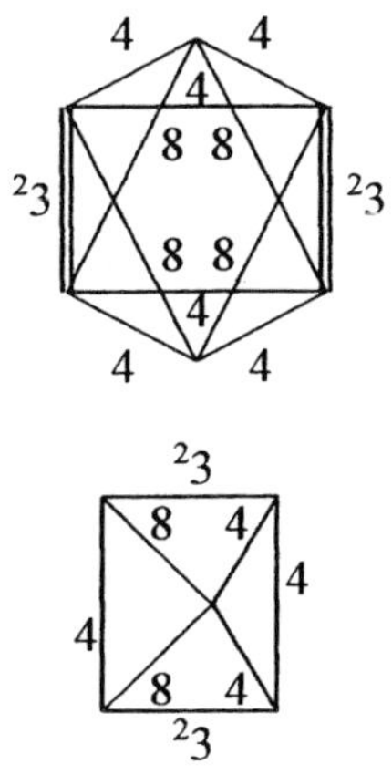

Cuboctacuboctahedral Closepack

Fig.38: Rhombicuboctahedral Closepacks

If we now try to subdivide the other triangle prism vertex figure, that of the rhombicuboctahedral closepack (p.13), we have a little more success with finding new arrangements not so far described.

Overleaf in Figure 38 are shown two closepacks with pyramidal vertex figures that can be found within this triangular prism vertex figure.

These in turn break down into four further tetrahedral vertex figure closepacks, two of which are 'hemi' forms shown opposite in Figure 40. The other two comprise the monoquasitruncated cuboctahedral closepack, already seen on p.25 and the truncated cuboctahexahedral closepack, shown below in Figure 39, a doubled up version of the truncated cuboctaprism closepack (p.15).

Solids	E/V	F/V	S/V
tr hexahexahedron	6	6 □	2
cube		1 ⧈	2
truncated square tiling		4 ○	2
Hemirhombicuboctahexahedral Closepack			
truncated cuboctahedron	6	4 □	2
triqtruncated cuboctahedron		1 ⟁	1
triquasitruncated octahedron		2 ⬡	1
truncated square tiling		4 ○	2
Hemicuboctacuboctahedral Closepack			
truncated cuboctahedron	6	4 □	4
truncated hexahexahedron		1 ⧈	2
		2 ⬡	
		4 ○	
Truncated Cuboctahexahedral Closepack			

Table 21: More Rhombicuboctahedrals

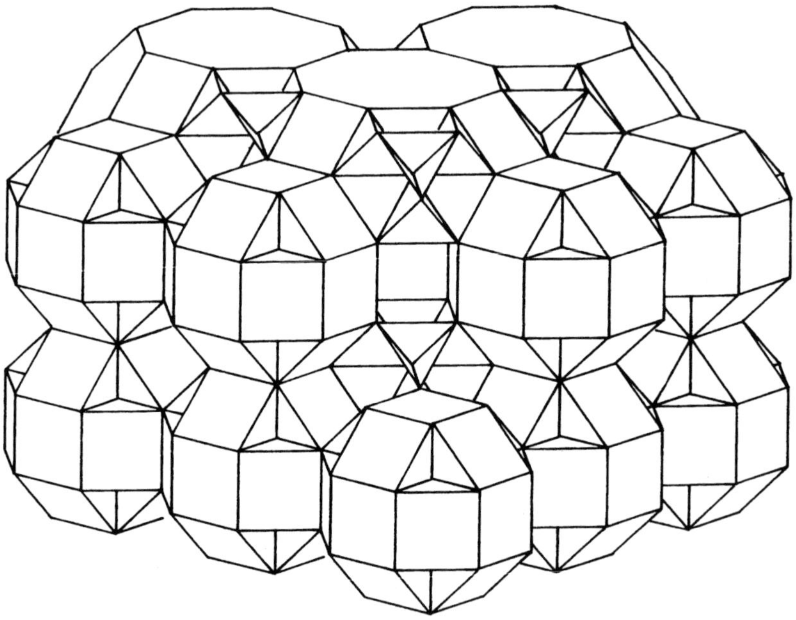

Truncated Cuboctahexahedral Closepack

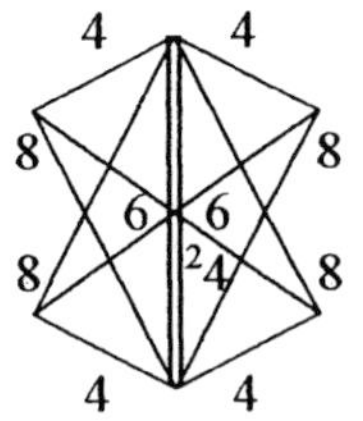

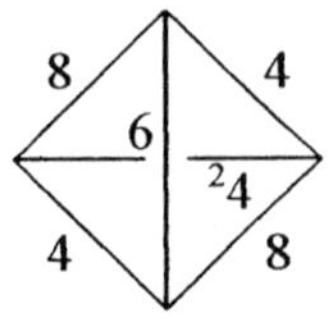

Fig.39: Rhombicuboctahedral Closepack

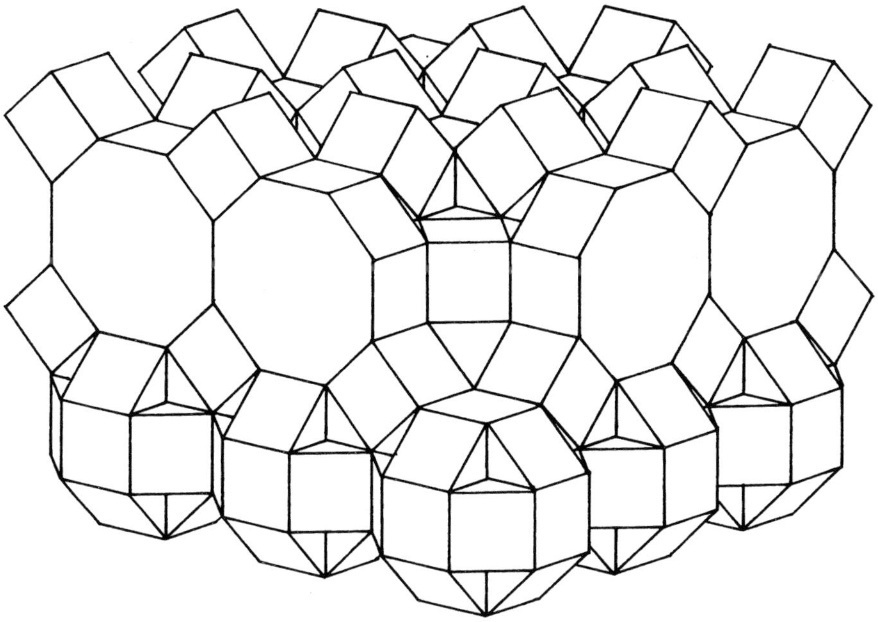

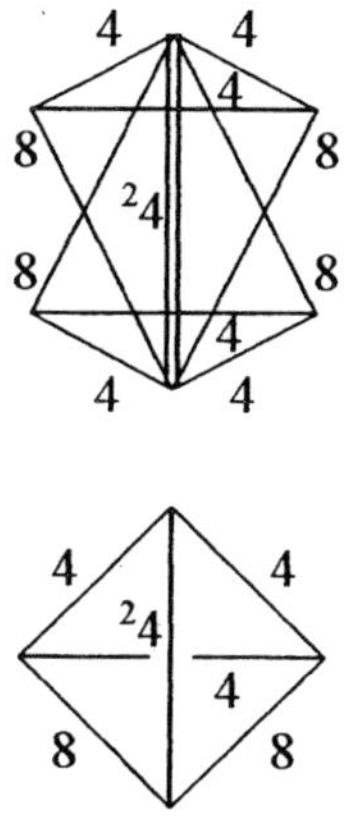

Hemirhombicuboctahexahedral Closepack

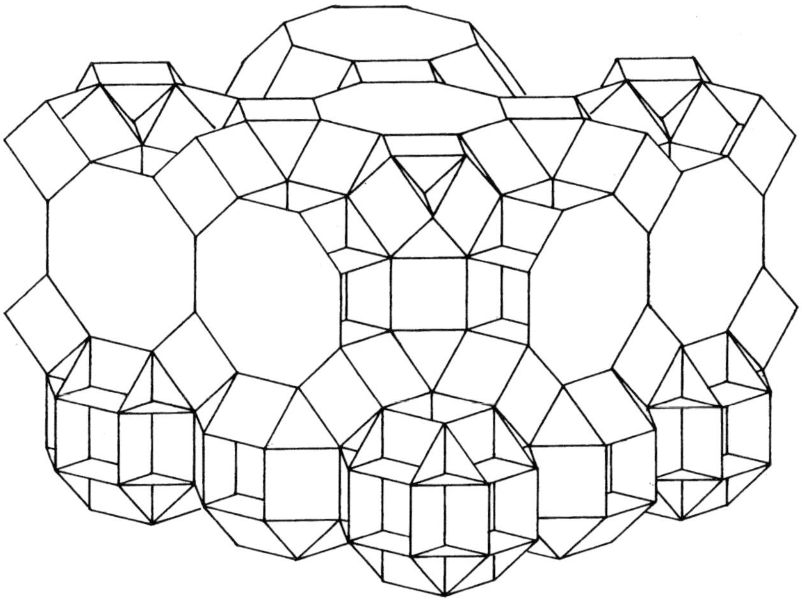

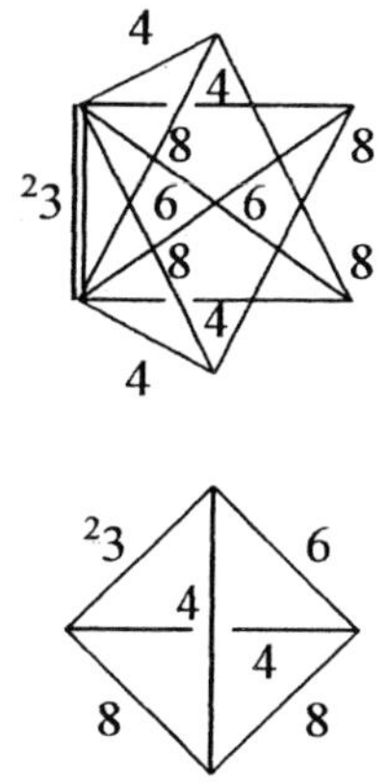

Hemicuboctacuboctahedral Closepack

Fig.40: Rhombicuboctahedral Closepacks

QUASIRHOMBICUBOCTETRAHEDRALS

A similar situation holds to that just seen if we try to subdivide the quasi forms of these same closepacks. Firstly we find that the quasirhombicuboctetrahedral closepack (p.24) can also be subdivided into one tetrahedral and one pyramidal vertex figure or alternatively at most three tetrahedral vertices.

As before we have already seen the quasiquasitruncated cuboctetrahedral closepack (p.22) and its other half the quasirhombicuboctagramprism closepack (p.33), the latter breaking down into the hemiquasirhombicuboctagramprism (p.41) and the hemiquasihexahexoctagramprism (p.43) closepacks just seen.

Again there is one with a similar vertex figure to this last, not seen before, which is shown below in Figure 41.

Solids	E/V	F/V	S/V
qrhombicuboctahedron	6	6 □	1
tetraqtruncated cuboctahedron		1 △	2
cube		1 ⧈	2
qtruncated hexahexahedron		2 ⬡	1
triquasitruncated octahedron		2 ✡	1
Quasirhombicuboctahexahedral Closepack			
cuboctahedron	6	6 □	1
qqtruncated cuboctahedron		2 △	2
cube		4 ✡	2
qqtruncated square tiling			2
Quasicuboctacuboctahedral Closepack			
quasitr hexahexahedron	6	3 □	3
tr stellated hexahedron		3 ⧈	1
qqtruncated square tiling		6 ✡	3
Hemiquasirhombicuboctetrahedral Cp			

Table 22: Quasirhombicuboct(etr)ahedrals

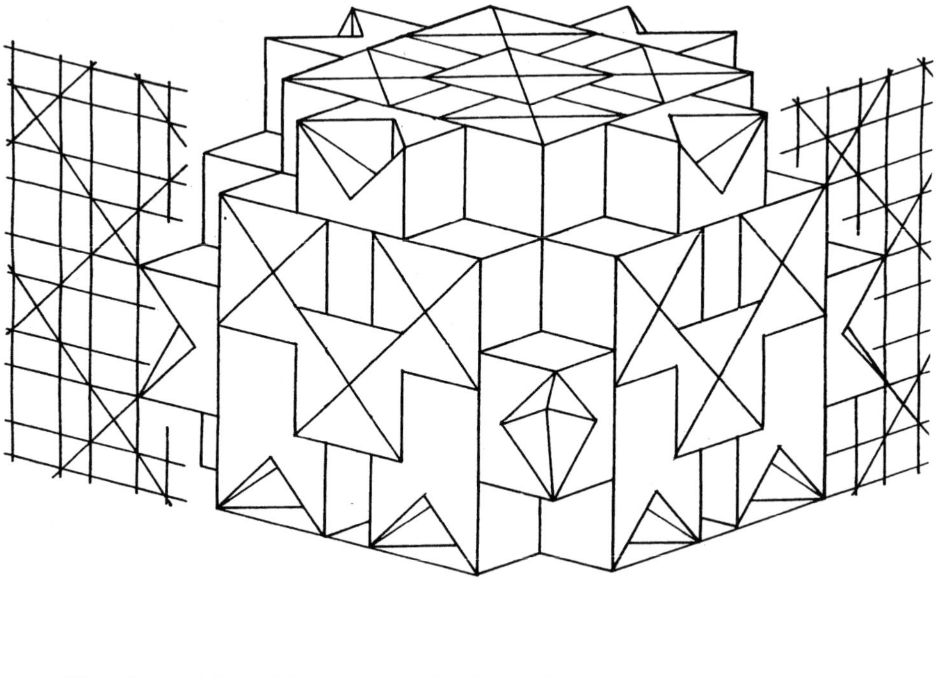

Hemiquasirhombicuboctetrahedral Closepack

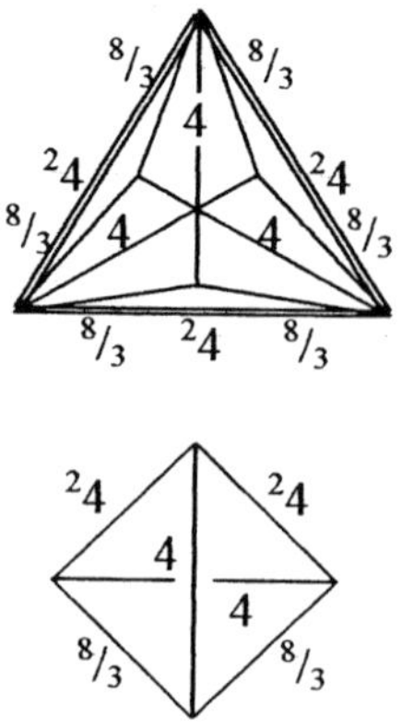

Fig.41: Quasirhombicuboctetrahedral Closepack

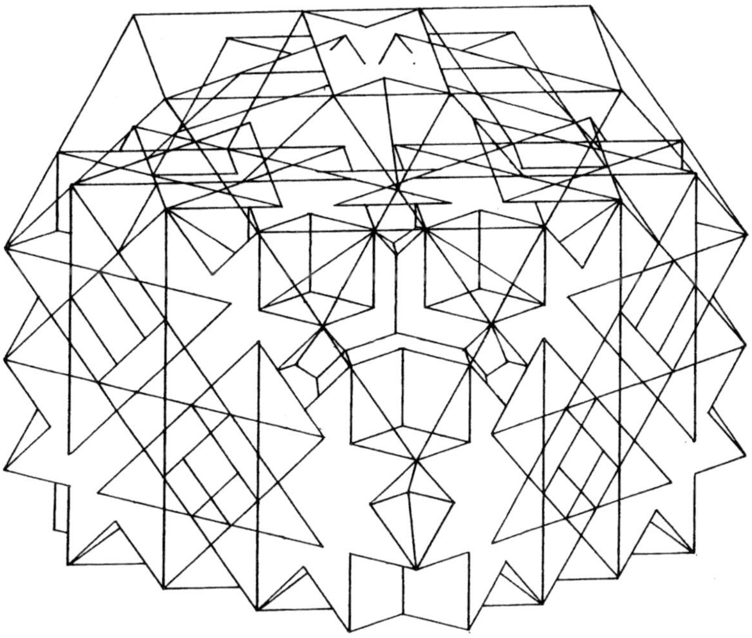

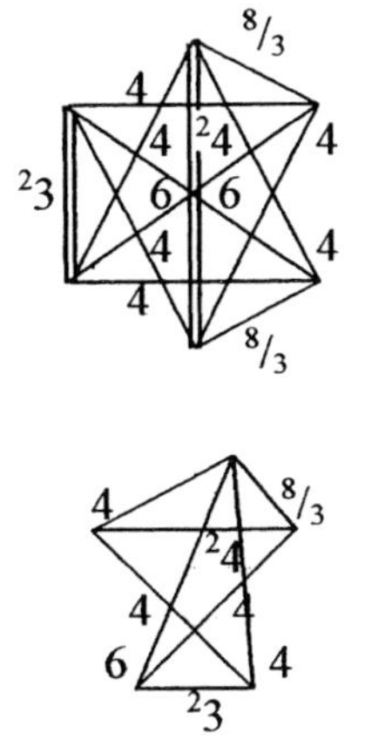

Quasirhombicuboctahexahedral Closepack

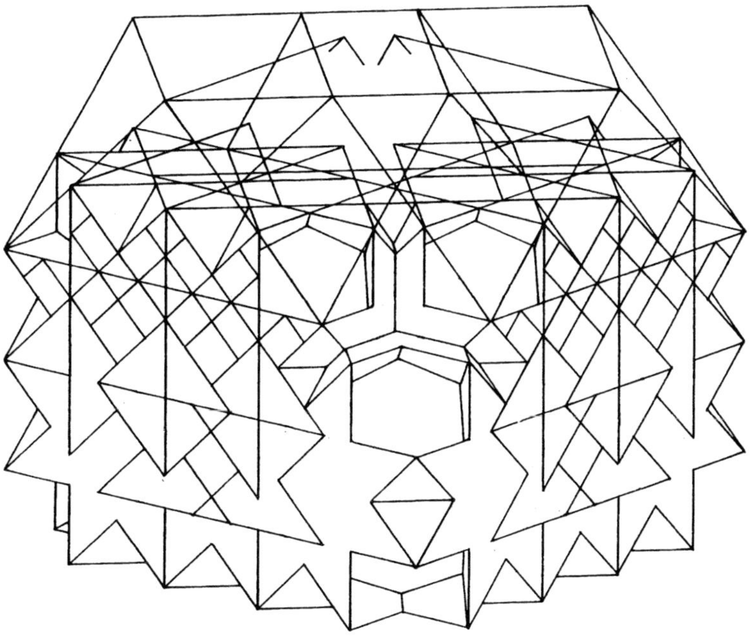

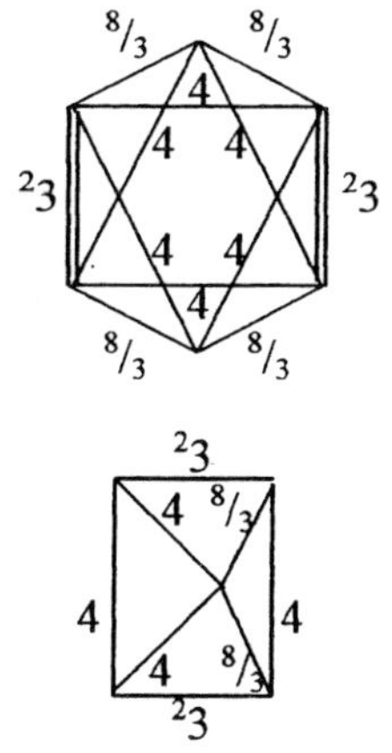

Quasicuboctacuboctahedral Closepack

Fig.42: Quasirhombicuboctahedral Closepacks

QUASIRHOMBICUBOCTAHEDRALS

As with the normal closepacks, if we now try to subdivide the other quasi-closepack triangle prism vertex figure, that of the quasirhombicuboctahedral closepack (p.28), we have a little more success with finding arrangements not so far described.

Overleaf in Figure 42 are shown the two closepacks with pyramidal vertex figures that can be found within this one.

These in turn break down into four further tetrahedral vertex figure closepacks, two of which are 'hemi' forms shown opposite in Figure 44. The other two again comprise one already seen, this time the diquasitruncated cuboctahedral closepack on p.25, and the tetraquasitruncated cuboctahexahedral closepack shown below in Figure 43, a doubled up version of the tetraquasitruncated cuboctaprism closepack (p.30).

Solids	E/V	F/V	S/V
quasitr hexahexahedron	6	6 □	2
cube		1 ⧈	2
qqtruncated square tiling		4 ✵	2
Hemiquasirhombicuboctahexahedral Cp			
tetraqtr cuboctahedron	6	4 □	2
qqtruncated cuboctahedron		1 △	1
triquasitr octahedron		2 ⬡	1
qqtruncated square tiling		4 ✵	2
Hemiquasicuboctacuboctahedral Closepack			
tetraqtr cuboctahedron	6	4 □	4
quasitr hexahexahedron		1 ⧈	2
		2 ⬡	
		4 ✵	
Tetraquasitruncated Cuboctahexahedral Cp			

Table 23: More Quasirhombicuboctahedrals

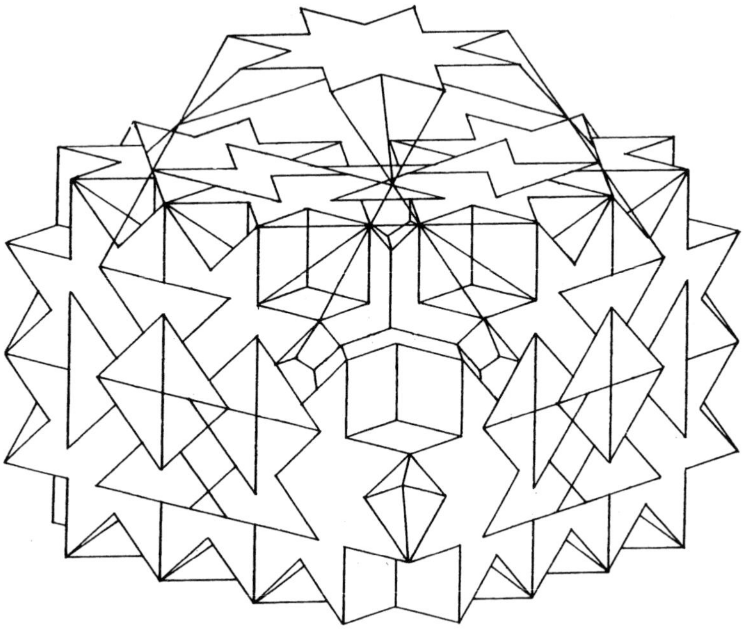

Tetraquasitruncated Cuboctahexahedral Closepack

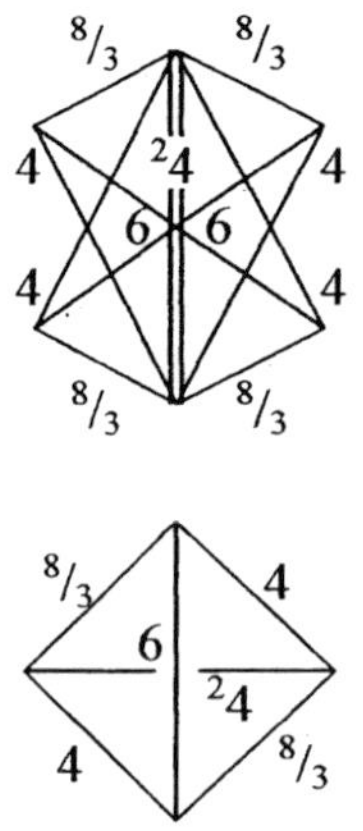

Fig.43: Quasirhombicuboctahedral Closepack

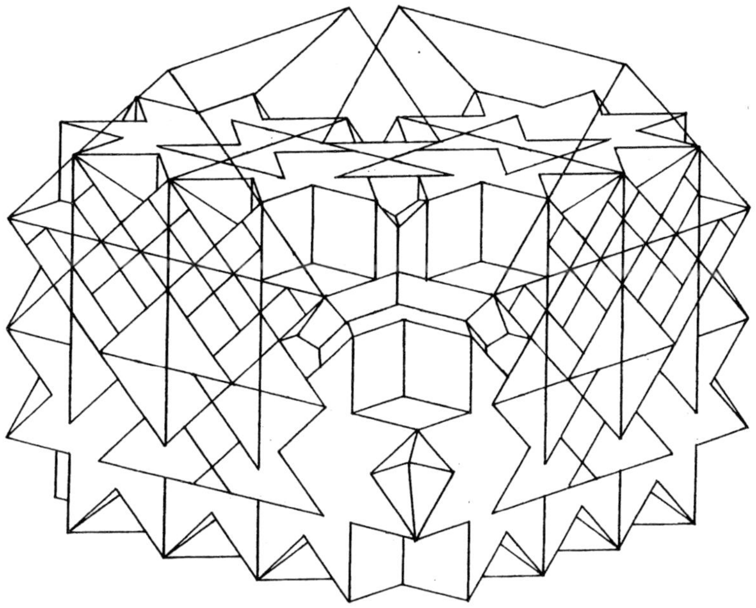

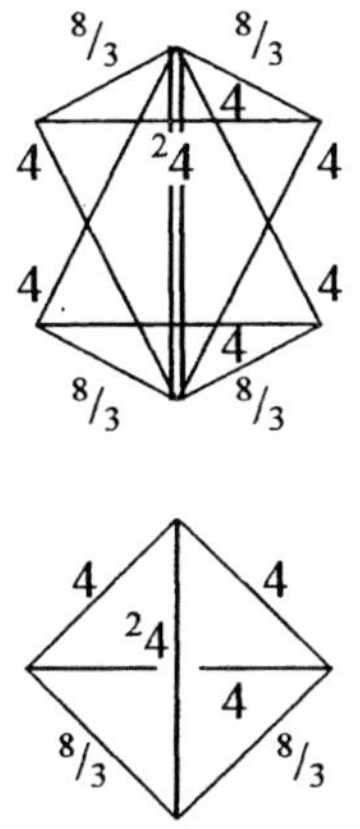

Hemiquasirhombicuboctahexahedral Closepack

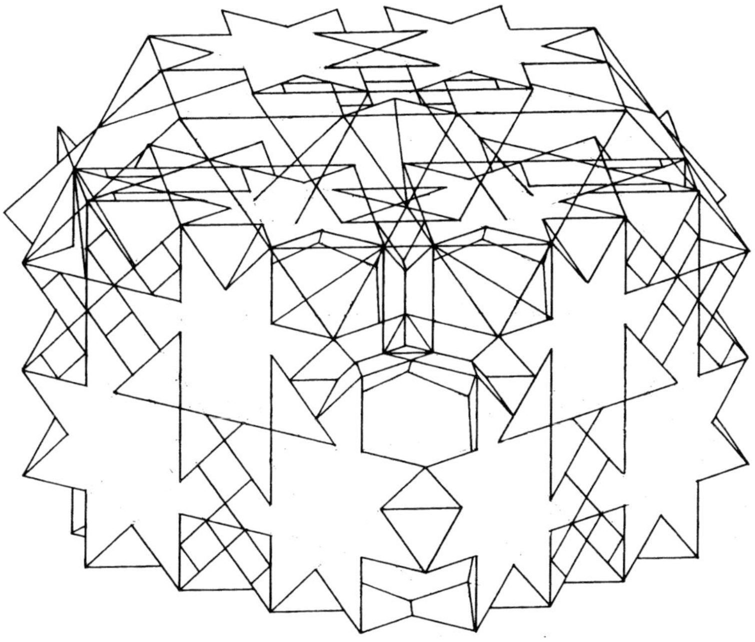

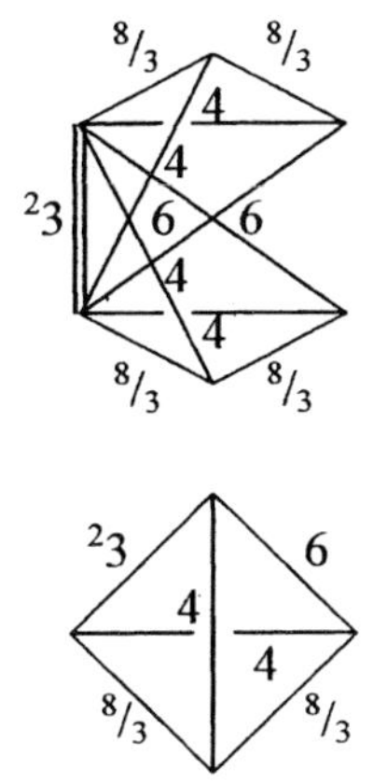

Hemiquasicuboctacuboctahedral Closepack

Fig.44: Quasirhombicuboctahedral Closepacks

We now come to the one remaining closepack vertex within which we have not yet sought further vertex figures. The cuboctahedral closepack has a cuboid vertex figure which can be cut in half in two different ways, each giving a triangular prism vertex figure closepack that includes a tiling as shown opposite in Figure 46.

The tetrahedral vertex figure of the quasi-truncated octahedral closepack (p.20) can be removed from each of these triangular prisms to leave a pyramidal vertex figure.

One of these gives rise to the full cuboid vertex figure again, complete with its internal faces, the other yields the multi-truncated octahedral closepack, with all three types of truncated octahedron present, as shown below in Figure 45.

Solids	E/V	F/V	S/V
cuboctahedron	8	4 △	2
octahedron		4 □	1
triquasitruncated octahedron		2 ◬	2
hexagon-triangle tiling		4 ⬡	2
Hemiquasicuboctahedral Closepack			
cuboctahedron	8	8 □	4
quasitruncated octahedron		8 ◬	2
square tiling			2
Hemicuboctahedral Closepack			
truncated octahedron	8	8 □	8
quasitruncated octahedron		4 ◬	1
triquasitruncated octahedron		8 ⬡	4
square tiling			2
Multitruncated Octahedral Closepack			

Table 24: Cuboctahedral Tiling Closepacks

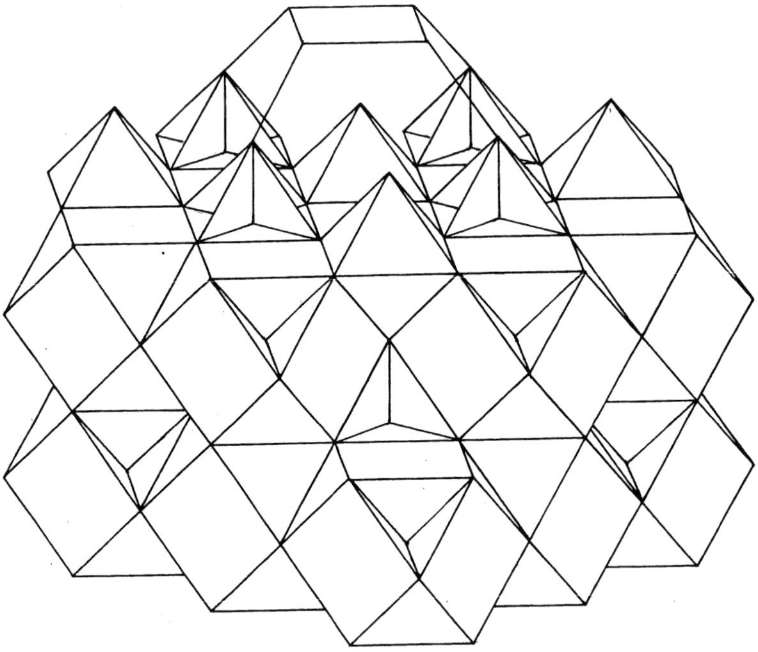

Multitruncated Octahedral Closepack

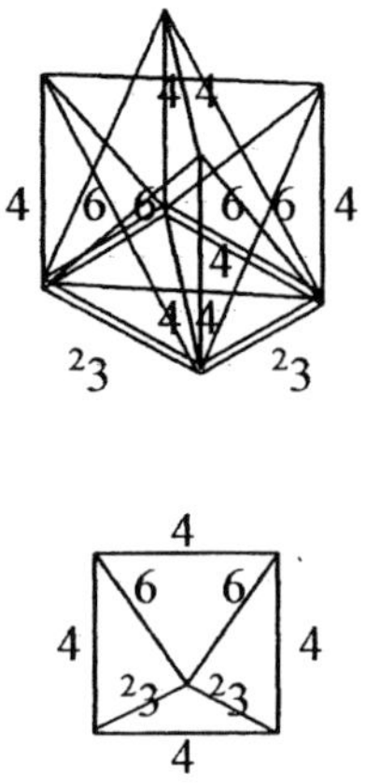

Fig.45: Cuboctahedral Tiling Closepack

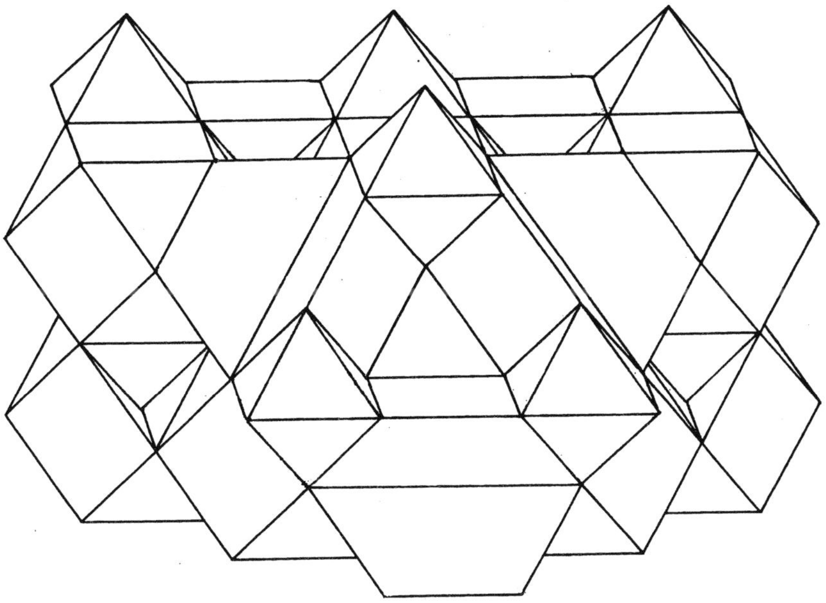

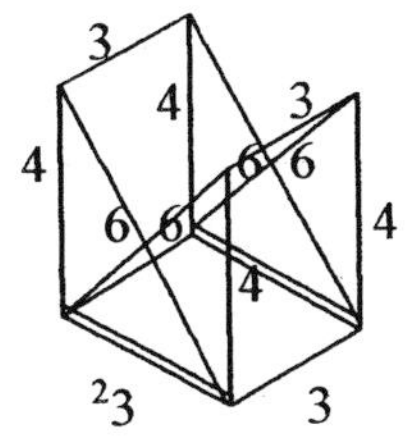

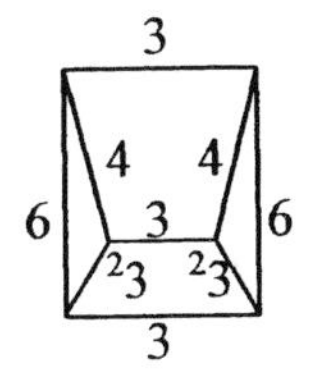

Hemiquasicuboctahedral Closepack

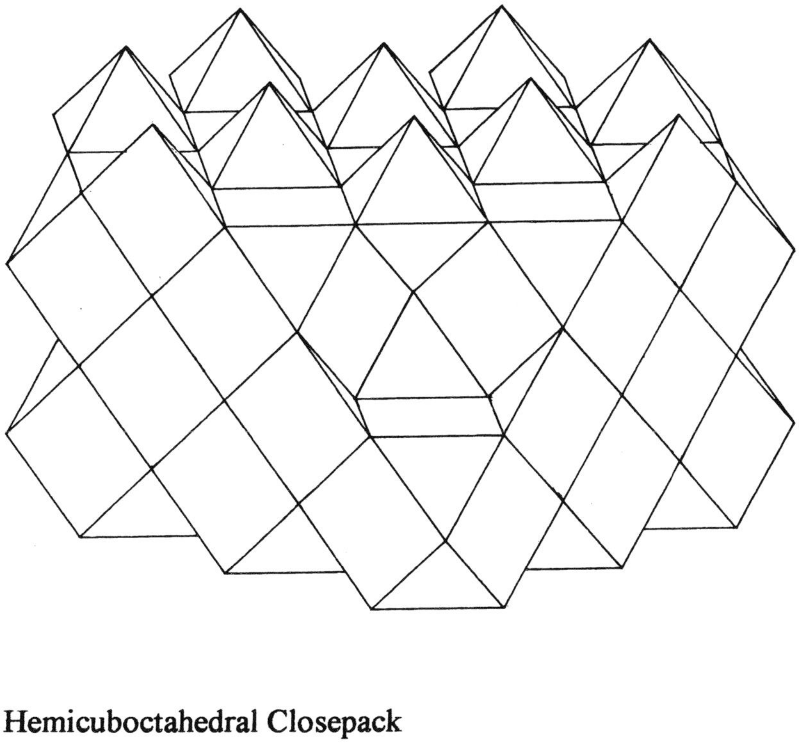

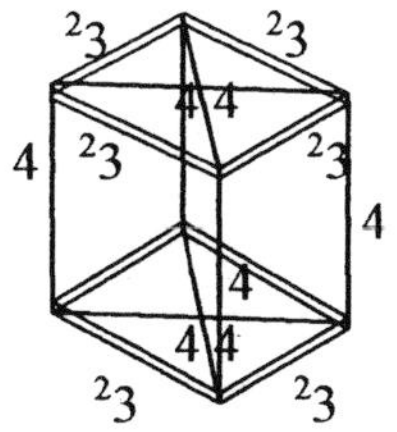

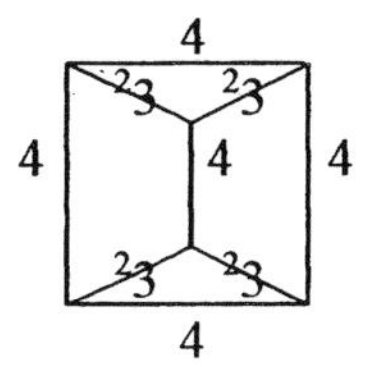

Hemicuboctahedral Closepack

Fig.46: Cuboctahedral Tiling Closepacks

We have now effectively dismantled all of the vertex figures seen thus far: pyramidal, triangular prism and cuboid, finding all the smaller hidden vertex figures within them.

Whilst dealing with the rhombicuboctahedral closepack and its 'quasi' equivalent, we encountered two closepacks that were doubled up versions of ones seen earlier. The truncated cuboctahexahedral (p.46) is a doubled up truncated cuboctaprism (p.15), whilst the tetraquasitruncated cuboctahexahedral (p.50) is a doubled up tetraquasitruncated cuboctaprism (p.30).

The two other non-degenerate truncated cuboctaprism type closepacks (p.29) yield similar double versions as shown opposite in Figure 48 and described in Table 25 adjoining.

Solids	E/V	F/V	S/V
truncated cuboctahedron	6	4 □	2
tetraqtruncated cuboctahedron		1 ◘	2
truncated hexahexahedron		2 ⬡	1
qtruncated hexahexahedron		2 ○	1
		2 ✵	
Monoquasitruncated Cuboctahexahedral Cp			
triqtr cuboctahedron	8	8 □	2
qqtruncated cuboctahedron		2 △	2
truncated hexahexahedron		2 ◘	2
qtruncated hexahexahedron		4 ○	2
		4 ✵	
Diquasitruncated Cuboctahexahedral Cp			
truncated cuboctahedron	6	4 □	2
tetraqtruncated cuboctahedron		2 △	2
triqtruncated cuboctahedron		2 ⬡	1
qqtruncated cuboctahedron		2 ○	1
hexagon-triangle tiling		2 ✵	1
Multitruncated Cuboctahedral Closepack			

Table 25: Double Cuboctaprism Closepacks

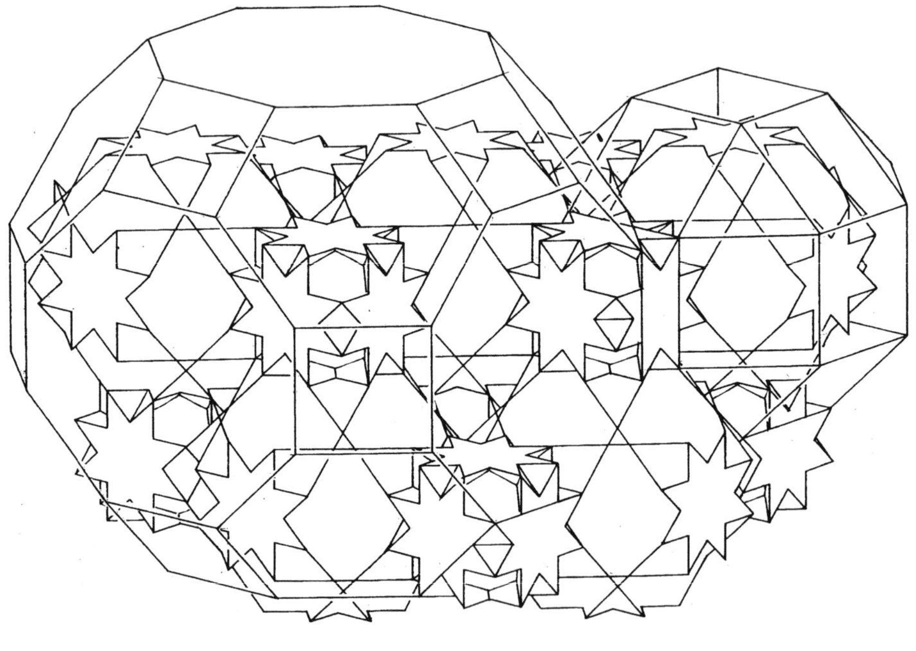

Multitruncated Cuboctahedral Closepack

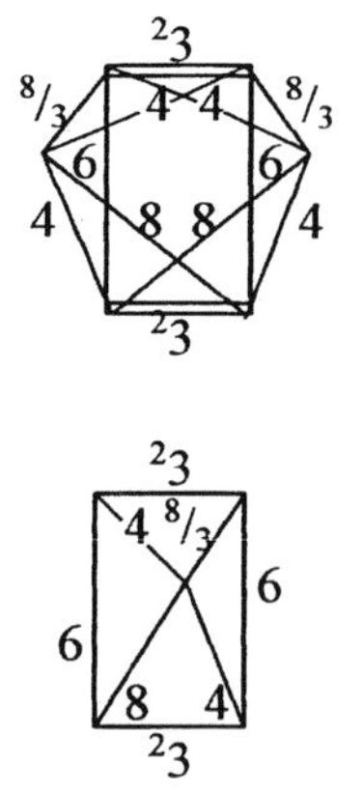

Fig.47: Another Triprism Vertex Closepack

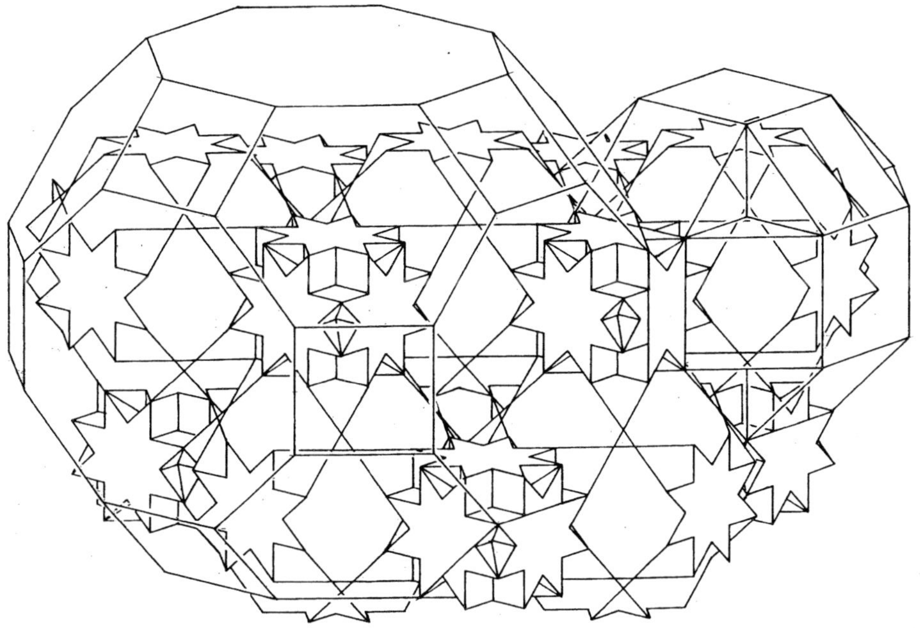

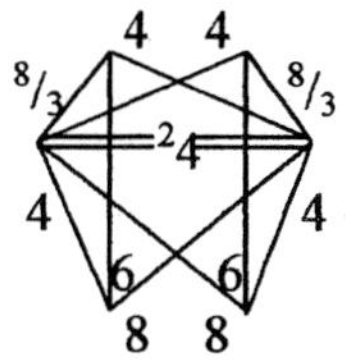

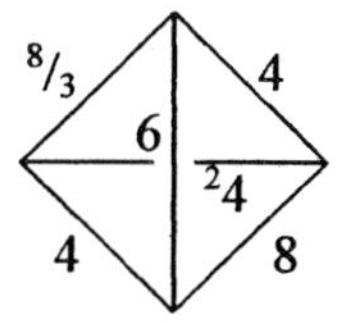

Monoquasitruncated Cuboctahexahedral Closepack

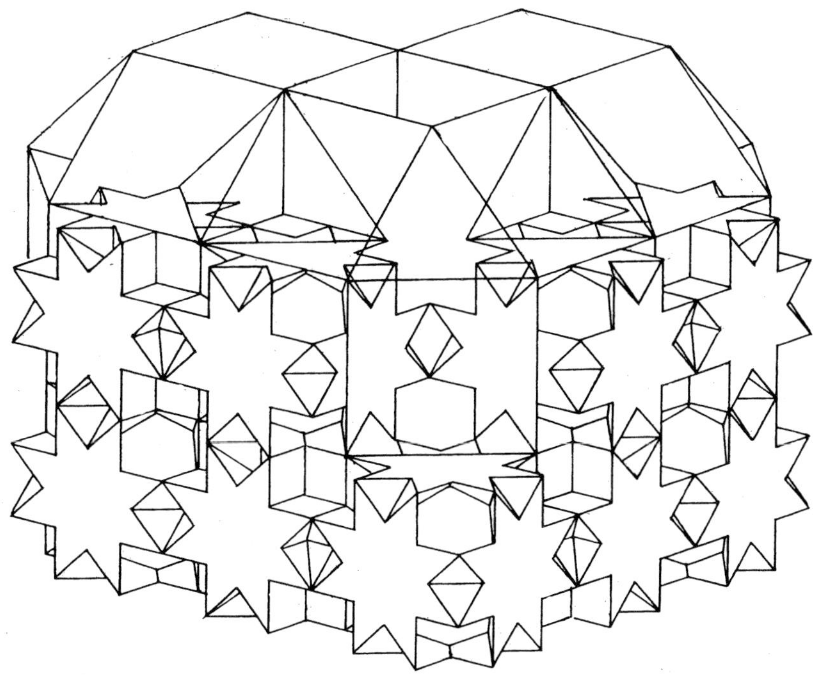

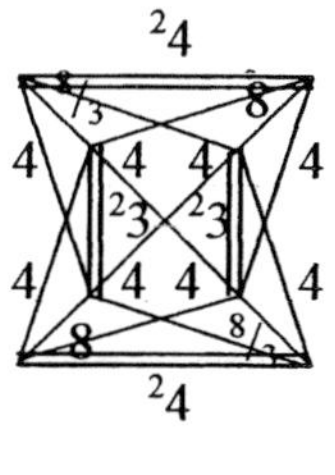

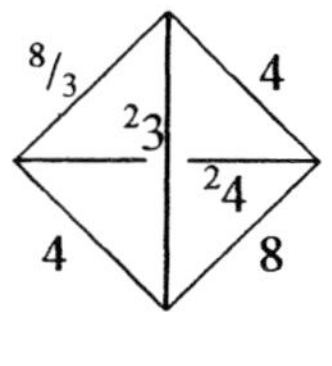

Diquasitruncated Cuboctahexahedral Closepack

Fig.48: Double Cuboctaprism Closepacks

ANOTHER CUBOID VERTEX

Overleaf above in Figure 48 the doubled up monoquasitruncated cuboctaprism closepack defines a further triangular prism vertex figure shell, with one face a hexagon-triangle tiling. Within this can be found the multitruncated cuboctahedral closepack with a pyramidal vertex figure, as shown in Figure 47. This contains all four types of truncated cuboctahedron in its make up.

Overleaf below in Figure 48 we have the resultant of doubling up the vertex figure of the diquasitruncated cuboctaprism closepack. This has further derivatives as shown below and opposite in Figures 49 and 50.

In all these instances of doubling up, an octagon / octagram prism has become a truncated / quasitruncated hexahexahedron with {4/2} symmetry.

Solids	E/V	F/V	S/V
tr hexahexahedron	8	8 □	4
quasitr hexahexahedron		4 ⧈	4
square tiling		4 ○	1
		4 ✵	
Hemihexaquasihexahedral Closepack			
qtruncated octahedron	8	10 □	1
triquasitr cuboctahedron		4 △	4
qqtruncated cuboctahedron		4 ○	4
qtruncated square tiling		4 ✵	4
Hemioctaquasicuboctahedral Closepack			
rhombicuboctahedron	8	4 △	4
quasirhombicuboctahedron		12 □	4
octahedron			1
square tiling			1
Rhombiquasirhombicuboctahedral Cp			

Table 26: More Cuboid Vertex Closepacks

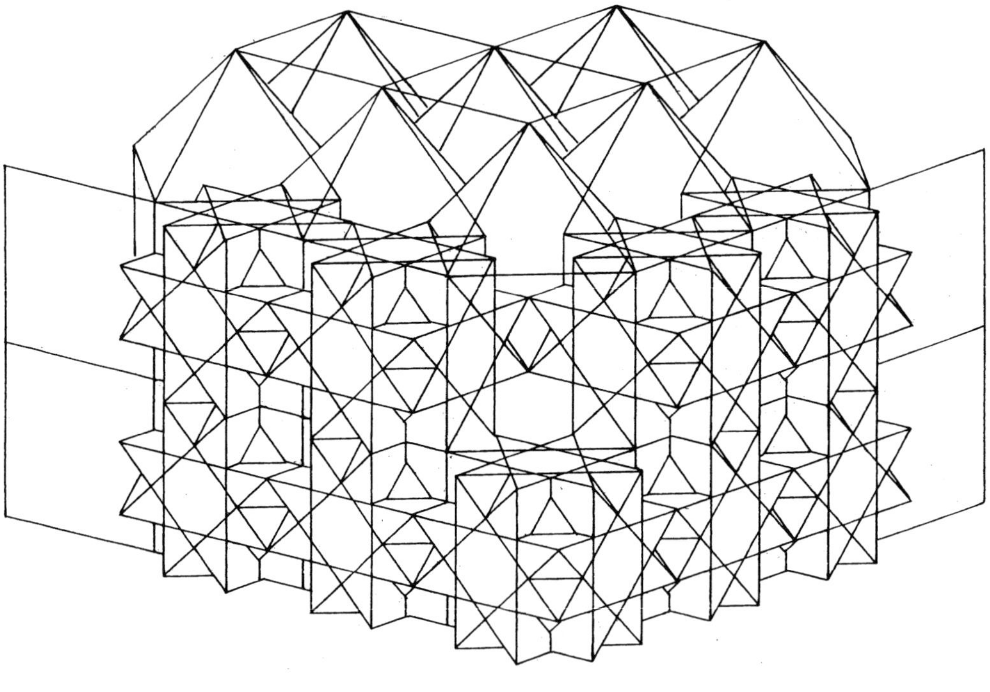

Rhombiquasirhombicuboctahedral Closepack

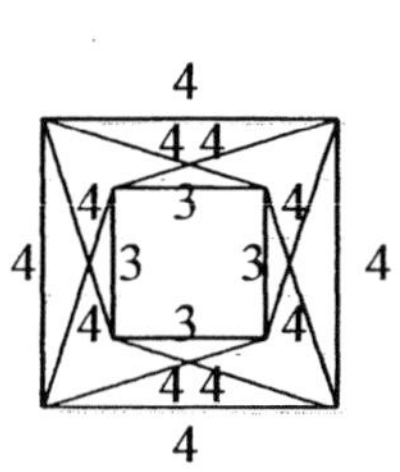

Fig.49: Another Cuboid Vertex Closepack

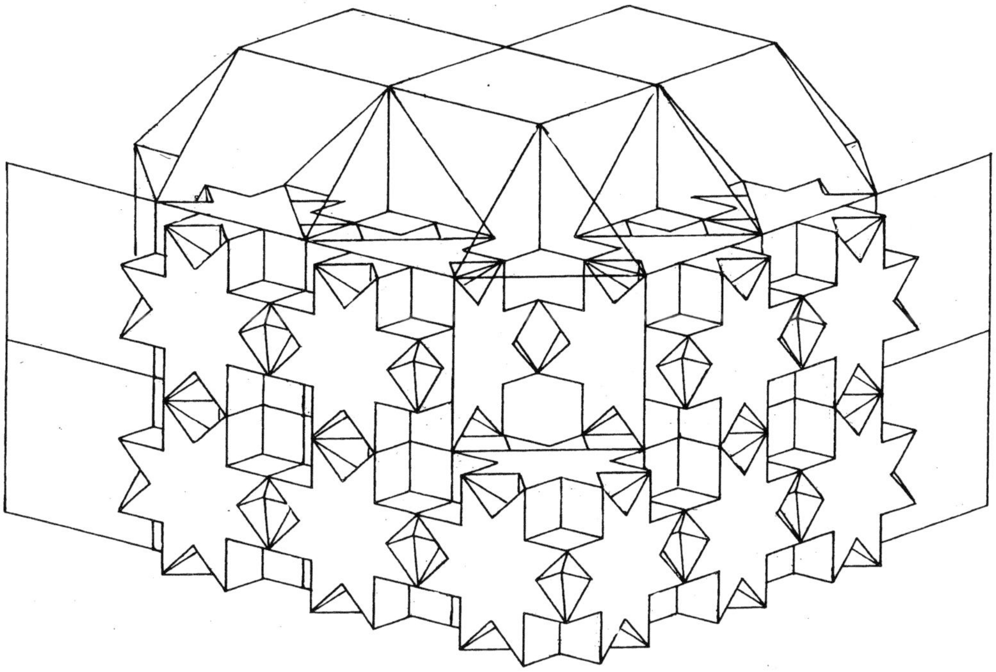

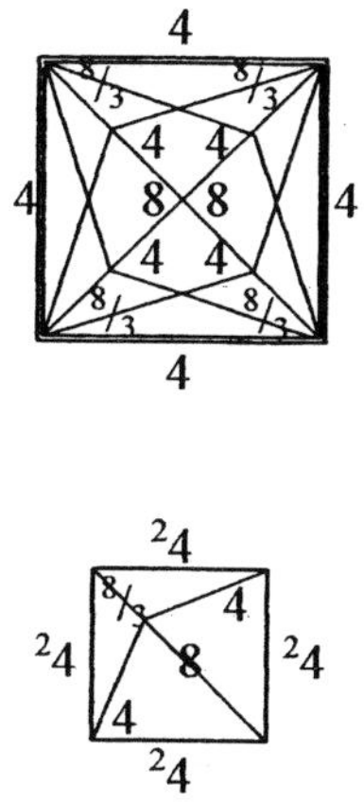

Hemihexaquasihexahedral Closepack

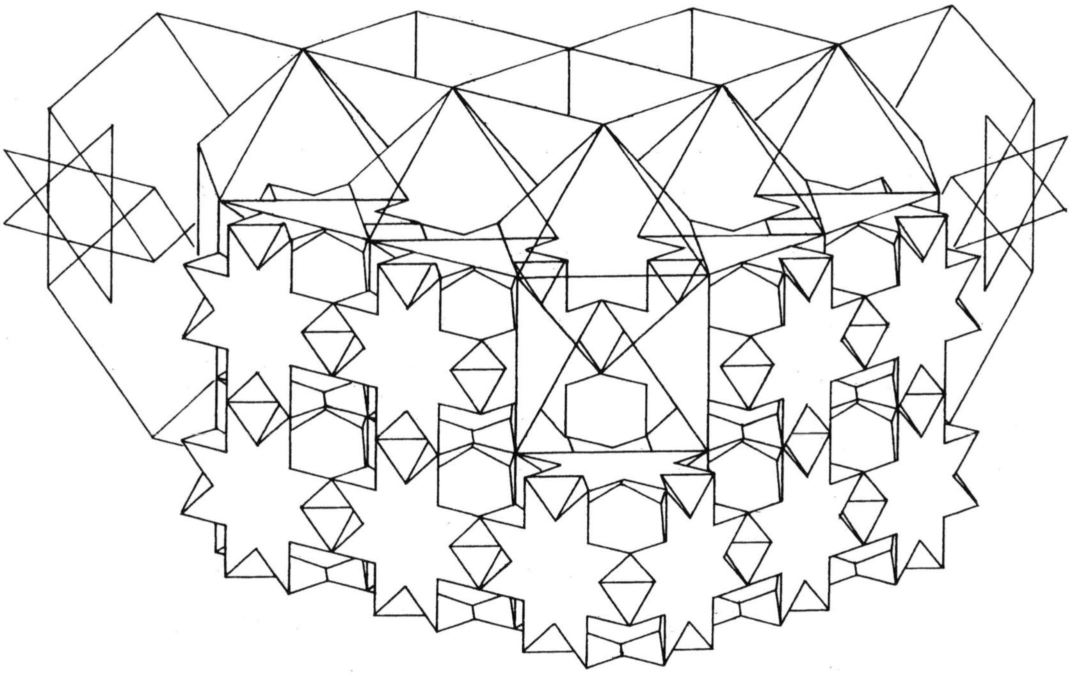

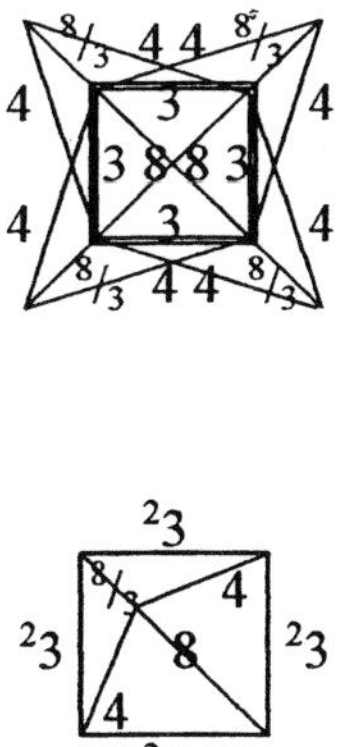

Hemioctaquasicuboctahedral Closepack

Fig.50: More Cuboid Vertex Closepacks

THE FINAL VERTEX SHELL

The shell of the vertex figure of the rhombiquasirhombicuboctahedral closepack and its close relations, seen overleaf, is also the shell of both the triquasitruncated and the quasiquasitruncated cuboctahedral closepacks seen earlier on page 27.

Reflecting this truncated pyramid shell in its base square tiling vertex gives a ten faced vertex figure with twelve vertices, as shown adjoining at the top of Figure 51. This can be seen to contain the smaller cuboid vertex figure of the cuboctahedral closepack, along with two crossed octahedral vertex figures of the cubic closepack we started with, as shown lower down in Figure 51.

We can in fact add a third crossed octahedral cubic closepack vertex figure to this arrangement, as shown at the top of Figure 52 opposite. This provides six more vertices to the twelve already there and allows three crossed cuboid cuboctahedral vertex figures to also be identified. These taken together form the cuboctahedral vertex figure of the octetrahedral closepack, the other 'regular' closepack started with.

The square faces of the octetrahedral vertex figure represent octahedra and taken with a fifth point to form a pyramidal vertex figure, we get a complementary pair, either a truncated cubic closepack employing the further fifth point, or a quasitruncated cubic closepack employing the nearer one.

In fact this monster vertex figure with eighteen vertices, as shown opposite in Figure 52, actually contains all the other vertex figures we have so far discussed in this volume. Having examined larger and larger vertices in the last few pages, it comes as something of a relief to find closure at last.

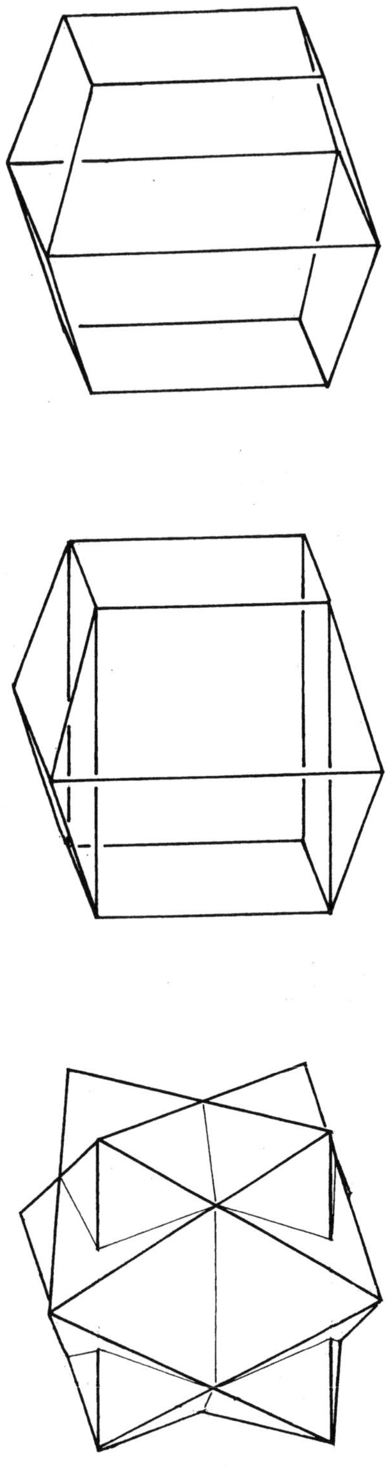

Fig.51: Assembling the Final Vertex

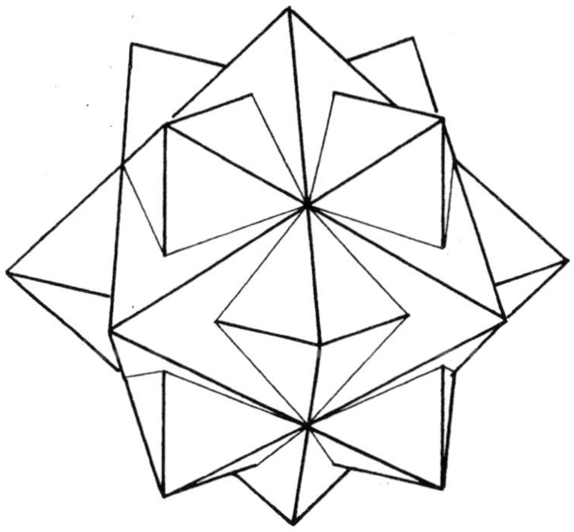

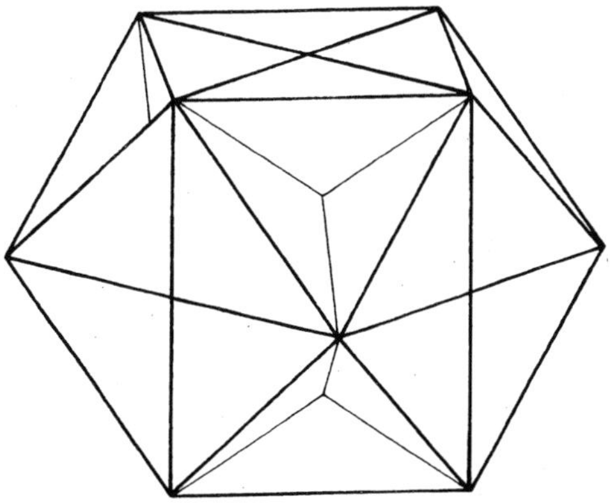

This 'final vertex figure' enclosing all the others seems to preclude any further discoveries. Overall it contains the twenty four triangle {3} faces of the octetrahedral closepack making up the basic cubocta-hedron framework. Each square face of this cuboctahedron has four octagrams {8/3} externally, mirrored by four octagons {8} internally, making a total of twenty four of each of these.

There are also forty eight square {4} faces present, twelve as diagonals of the cubocta-hedron's squares, thirty six internally, making up four octahedral cubic closepack vertices. Finally there are twenty four hexagon {6} faces present, six between each of four pairs of opposite cubocta-hedron triangles, forming the vertices of four odd closepacks.

In total this is 144 potential faces meeting at a single vertex! Hopefully we have covered all possible combinations of these making workable smaller vertex figures.

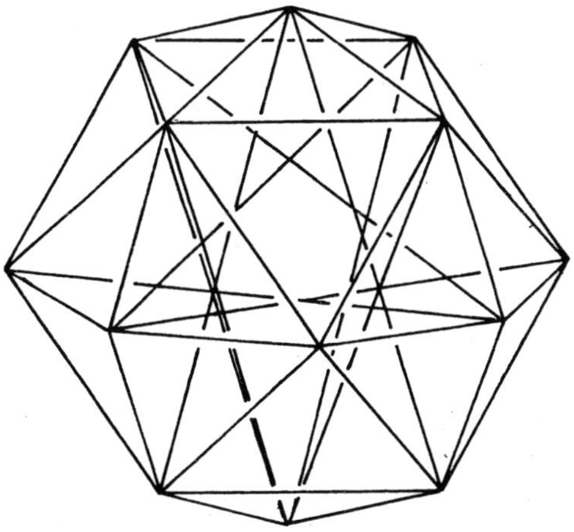

Fig.52: Within the Final Vertex

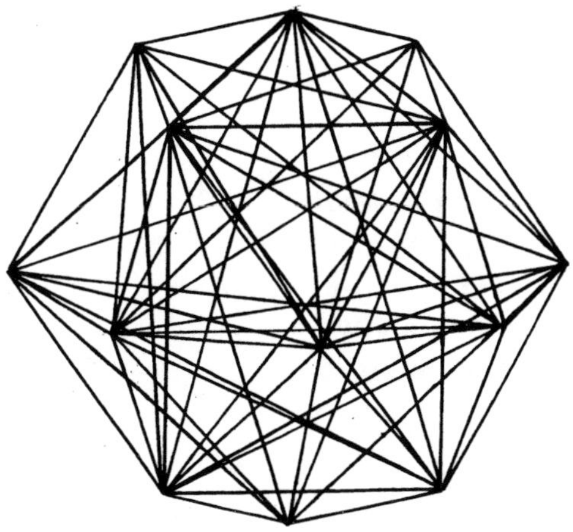

The Final Vertex

SUMMARY OF CLOSEPACKS

The fourteen (thirteen if we exclude the rhombic dodecahedron for not being regular) true 'uniform' closepacks are generally all that ever get described as closepacks.

We have hopefully now shown here that it is fruitful to expand our definition to include the 'quasi-closepacks', whose properties have been seen to closely shadow those of the 'true' ones. As with the quasi-truncated tilings, we have had to allow an element of interpenetration to occur in order to construct these. However this happens on an orderly basis and has not upset the regularity of the results obtained.

As a general rule all the 'true closepack' vertex figures from pages 6 through to 12 have the property of enclosing the centre point of the monster 'final vertex figure' within them. Conversely all the 'quasi-closepack' vertex figures from pages 18 through to 33 specifically do not include that centre point.

The 'hemis' that follow do include the centre point, but at the centre of a tiling vertex, on the surface of the vertex figure, whilst the 'hemi-quasis' have the centre point in the same plane as the tiling vertex, but outside of it.

It is perhaps significant that the inclusion of truncated polyhedra with double polygon faces $\{^{2}3\}$ and $\{^{2}4\}$, the result of truncating the even denominator polygons $\{3/2\}$ and $\{4/2\}$, does not upset the apple-cart either. The results have remained well-behaved and we have achieved closure by obtaining the 'final vertex figure', that which encloses and includes all the others we have discussed.

We will now take a quick overview of the closepacks as but one layer of many amongst what are termed the 'honeycombs' before finishing our survey with an appendix looking at the 'tiling closepacks'.

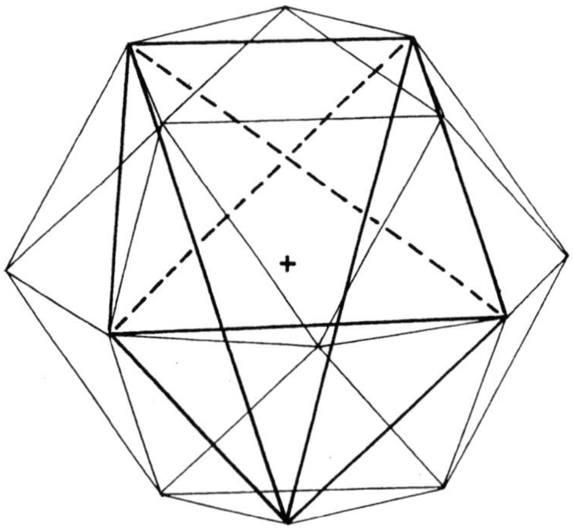

Rhombicuboctahedral (+Hemis)

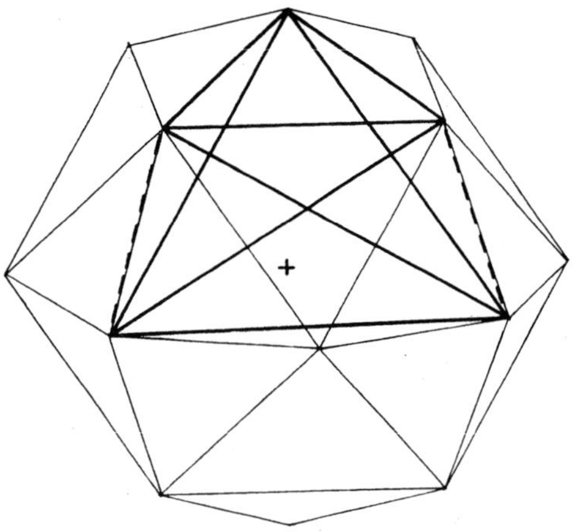

Quasirhombicuboctahedral (+Hemis)

Fig.53: True, Quasi and Hemi Vertices

A NOTE ON REGULAR HONEYCOMBS

In the second volume of this series *'Incomplete Tilings'* we examined assemblies of polygons that filled the plane. In the first part of this volume we saw assemblies of polyhedra that filled three dimensions, the true closepacks.

These form the first parts of a more general picture in higher dimensions. The general 'honeycomb' that fills n dimensions with n-dimensional figures is known as δ_{n+1} and has a formula $\{4,3^{n-2},4\}$. The 'n+1' acknowledges the relationship of such figures to the next dimension up, where n dimensions are folded up to yield (n+1)-dimensional figures. Thus the tilings of the plane are related to the polyhedra (effectively tilings of the sphere) and in their turn the closepacks will be seen to be related to the hypersolids.

This series of space filling polytopes thus begins with δ_3, the square tiling {4,4} that fills two dimensions, continues with δ_4, the cube closepack {4,3,4} that fills three dimensions, and goes on to δ_5, {4,3,3,4}, an assembly of hypercubes that fills four dimensions, then δ_6, {4,3,3,3,4} that fills five dimensions etc.

These are interestingly all self inverses, whilst the only other regular honeycombs seem to come as inverse pairs. These comprise the triangle and hexagon tilings, {3,6} and {6,3}, that fill two dimensions, and another pair {3,3,4,3} and {3,4,3,3} that fill four dimensions with hyperocta-tetrahedra {3,3,4} and hyperoctahedra {3,4,3} respectively. Both of these dual pairs have some affinity with our octetrahedral closepack that filled three dimensions with octahedra and tetrahedra, however, there do not appear to be any known higher analogues.

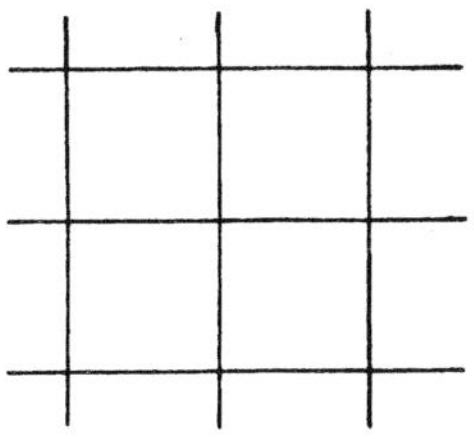

2D δ_3 {4,4} square tiling

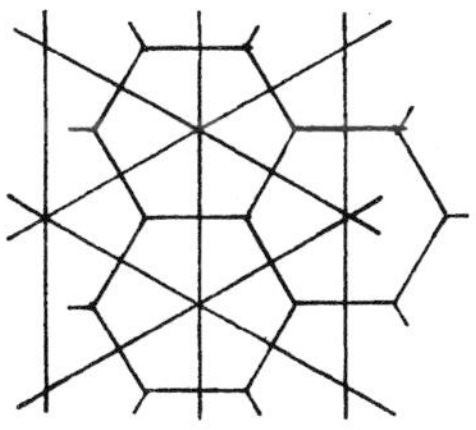

Plus {3,6} & {6,3}

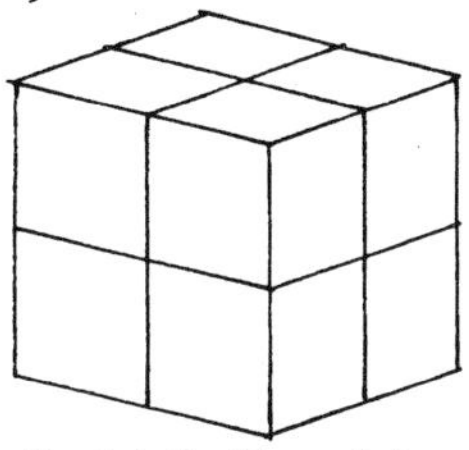

3D δ_4 {4,3,4} cubic closepack

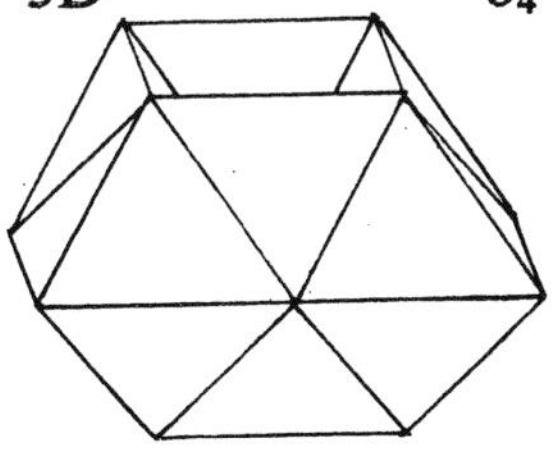

Plus {octet} closepack

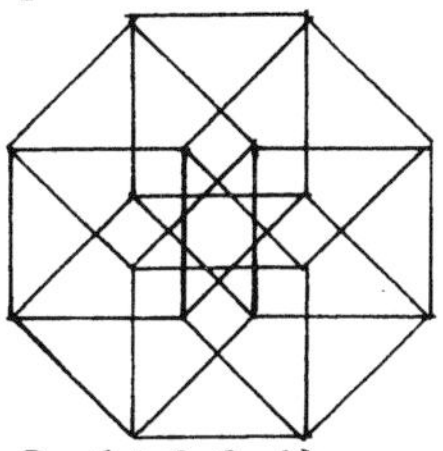

4D δ_5 {4,3,3,4}

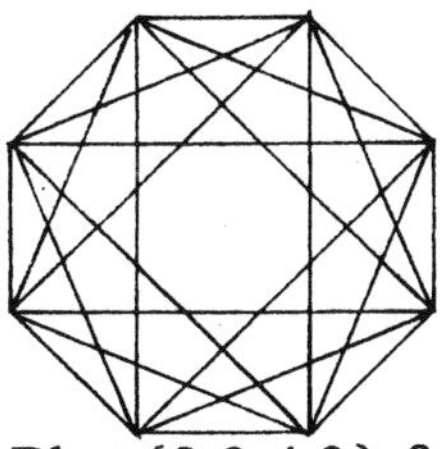

Plus {3,3,4,3} & {3,4,3,3}

Fig.54: Regular Honeycombs

APPENDIX: TILING CLOSEPACKS

The tilings that fill the plane in two dimensions, can in fact provide us with some further, albeit simple, vertex figures.

Any polygonal tiling in two dimensions can easily be made three dimensional by treating the polygons as prisms to form layers one edge length thick extending to infinity in two dimensions.

The simple process of stacking these layers gives us the tiling closepacks. They all have dipyramidal vertex figures formed by adding square faces both sides of the original tiling's polygonal vertex figure.

The simplest of these is that produced from the square tiling, yielding the octahedral vertex figure of the cubic closepack as shown below in Figure 55.

Solids	E/V	F/V	S/V
hexaprism	5	3 ⬡ 6 □	6
Hexagon Tiling Closepack			
triprism	8	6 △ 12 □	12
Triangle Tiling Closepack			
square prism (cube)	6	4 □ 8 □	8
Square Tiling (Cubic) Closepack			

Table 27: Regular Tiling Closepacks

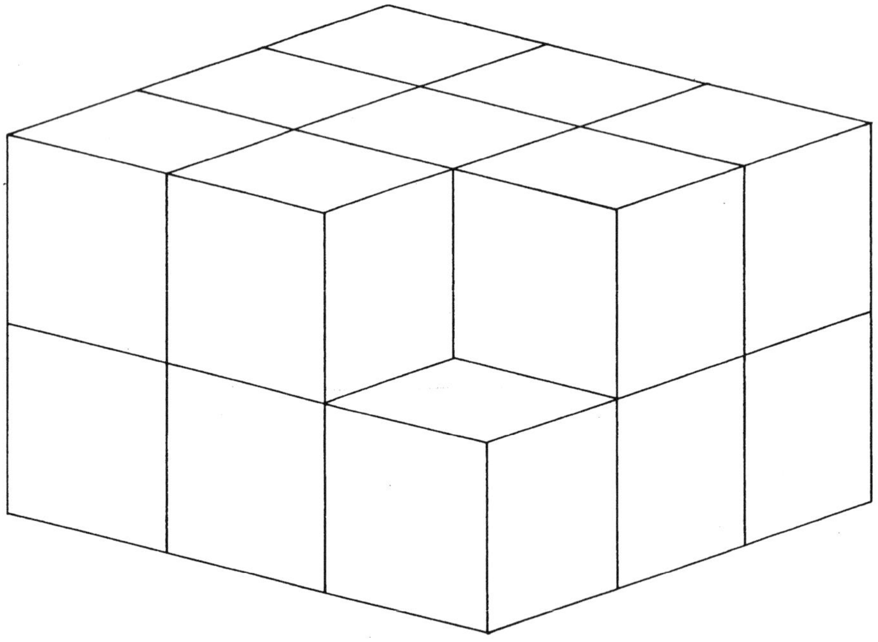

Square Tiling (Cubic) Closepack

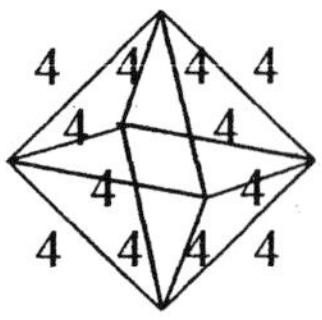

Fig.55: Square Tiling Closepack

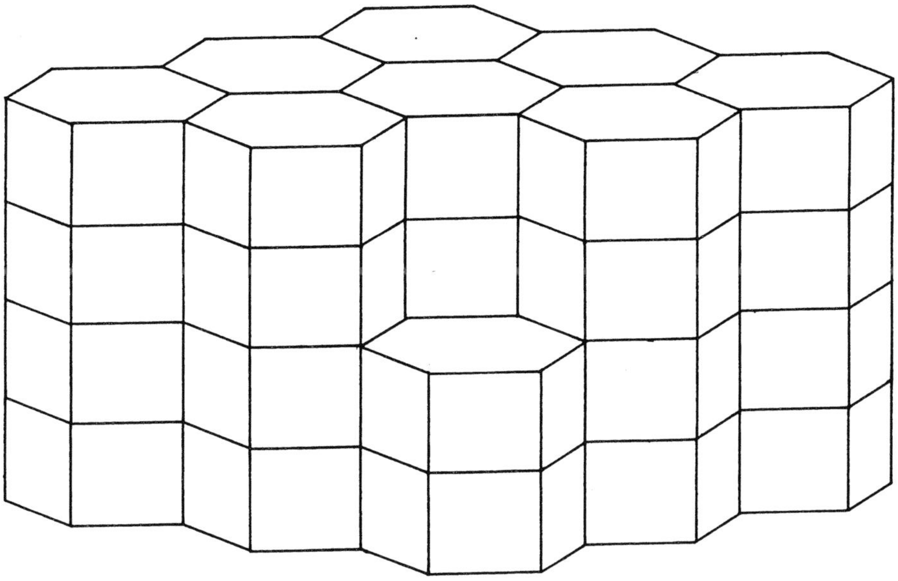

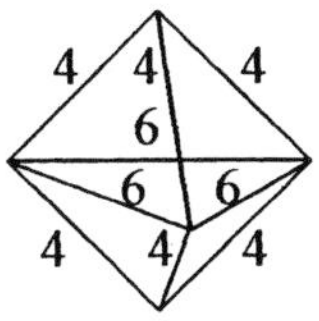

Hexagon Tiling Closepack

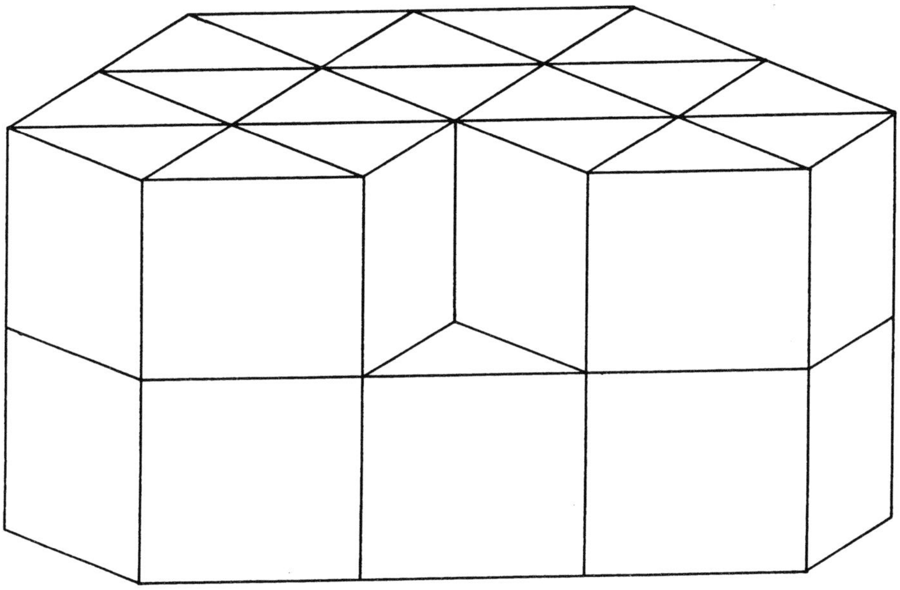

Triangle Tiling Closepack

Fig.56: Hexagon & Triangle Tiling Closepacks

TRUNCATED TILING CLOSEPACKS

Overleaf in Figure 56 are shown the other two regular tiling closepacks, produced using the hexagonal prism and triangular prism respectively.

The truncated square tiling (octagon-square) and truncated hexagon tiling (dodecagon-triangle) yield similar results when replaced by prisms, as shown opposite in Figure 58.

The truncated triangle tiling, however is simply the hexagon tiling again, but its second truncation is the hexagon-triangle tiling, which has an associated closepack as shown below in Figure 57. This is also the second truncation of the hexagon tiling, the quasi-regular form occurring at the centre of the truncation series.

Solids	E/V	F/V	S/V
octaprism	5	1 □	4
cube		2 ⯃	2
		6 □	
Truncated Square Tiling Closepack			
dodecaprism	5	1 △	4
triprism		2 ○	2
		6 □	
Truncated Hexagon Tiling Closepack			
hexaprism	6	2 △	4
triprism		2 ⬡	4
		8 □	
Hexagon-Triangle Tiling Closepack			

Table 28: Truncated Tiling Closepacks

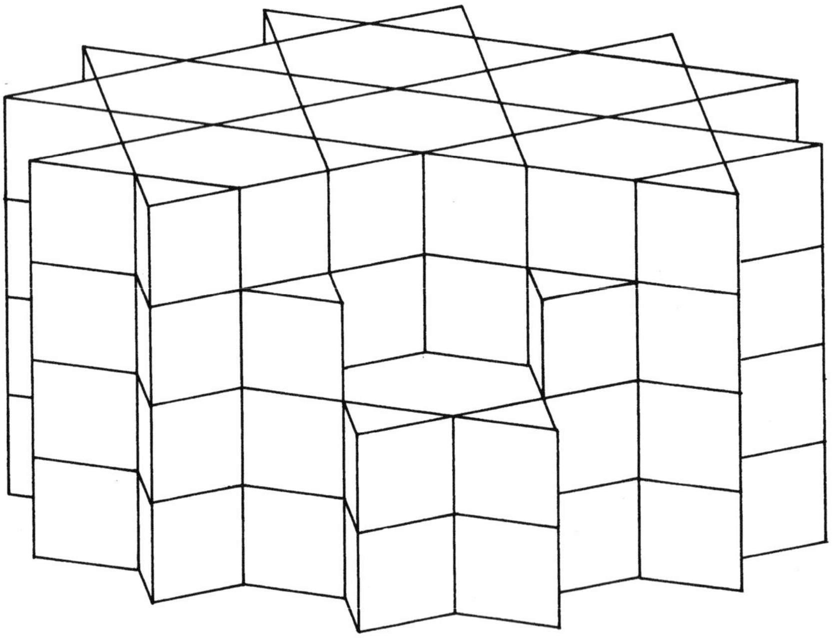

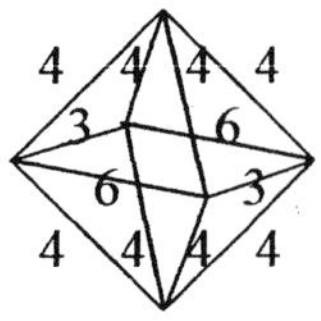

Hexagon-Triangle Tiling Closepack

Fig.57: Quasi-Regular Tiling Closepack

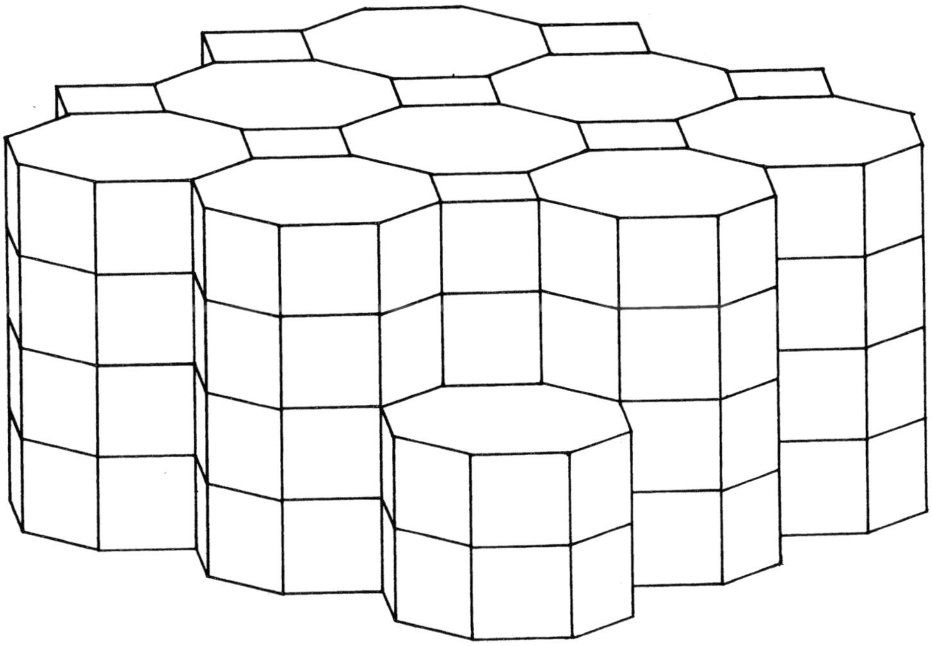

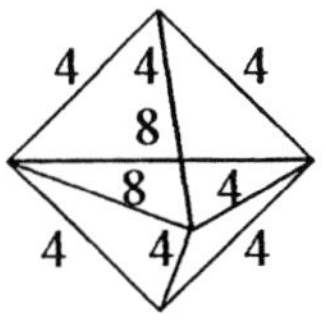

Truncated Square Tiling Closepack

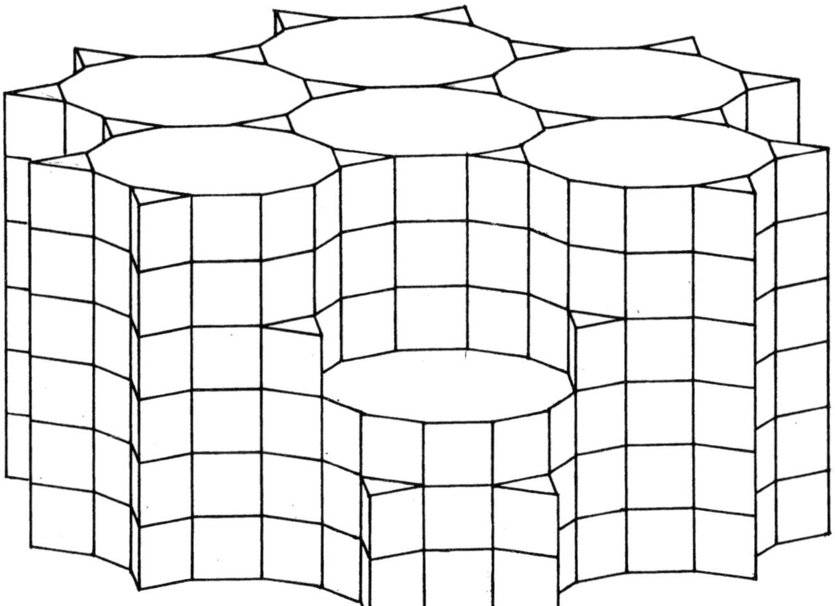

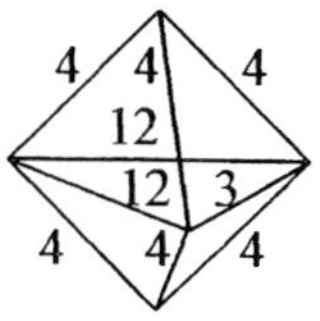

Truncated Hexagon Tiling Closepack

Fig.58: Truncated Tiling Closepacks

SNUB & ODD TILING CLOSEPACKS

All of the tiling closepacks so far discussed have had only one or two types of prism in their make up. Three other tiling closepacks have only two types of prism present in each.

The odd tiling closepack shown below in Figure 59 is a curious mixture of square and triangle tiling closepack, with two sets of parallel tiling planes perpendicular to each other.

Opposite in Figure 60 are shown the two snub tiling closepacks, that above for the square tiling family {4,4}, that below for the hexagon-triangle tiling family {3,6}. The former can be seen to be simply a rearrangement of the odd tiling closepack.

Solids	E/V	F/V	S/V
cube triprism	7	3 △ 2 □ 10 □	4 6
Snub Square Tiling Closepack			
hexaprism triprism	7	4 △ 1 ⬡ 10 □	2 8
Snub Hexagon Tiling Closepack			
cube triprism	7	3 △ 2 □ 10 □	4 6
Odd Tiling Closepack			

Table 29: Snub & Odd Tiling Closepacks

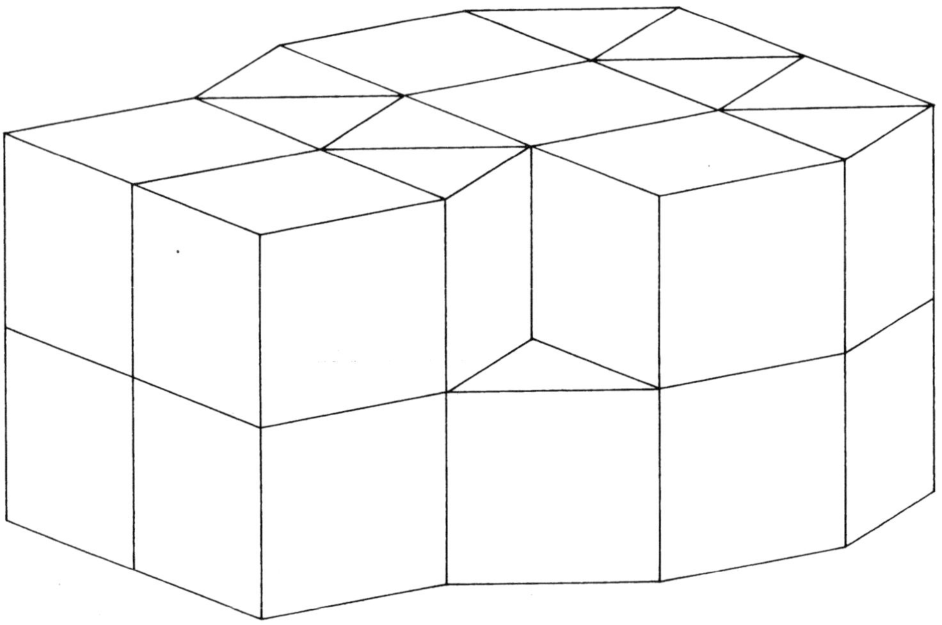

Odd Tiling Closepack

Fig.59: Odd Tiling Closepack

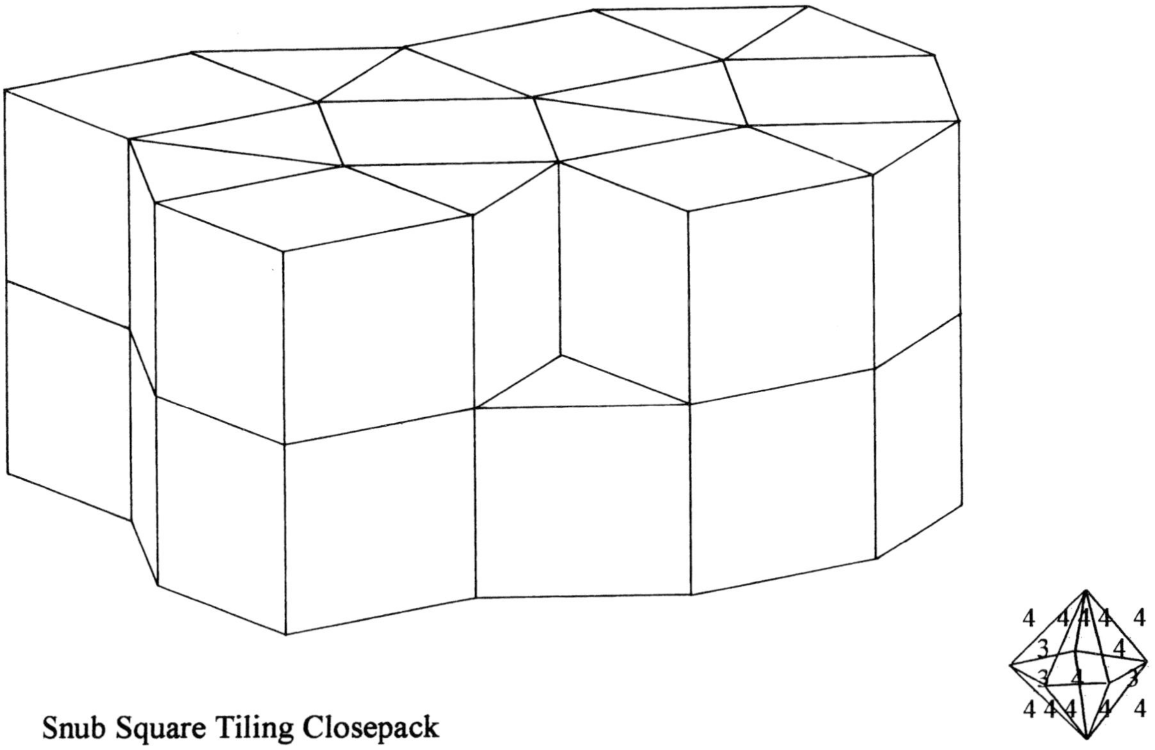

Snub Square Tiling Closepack

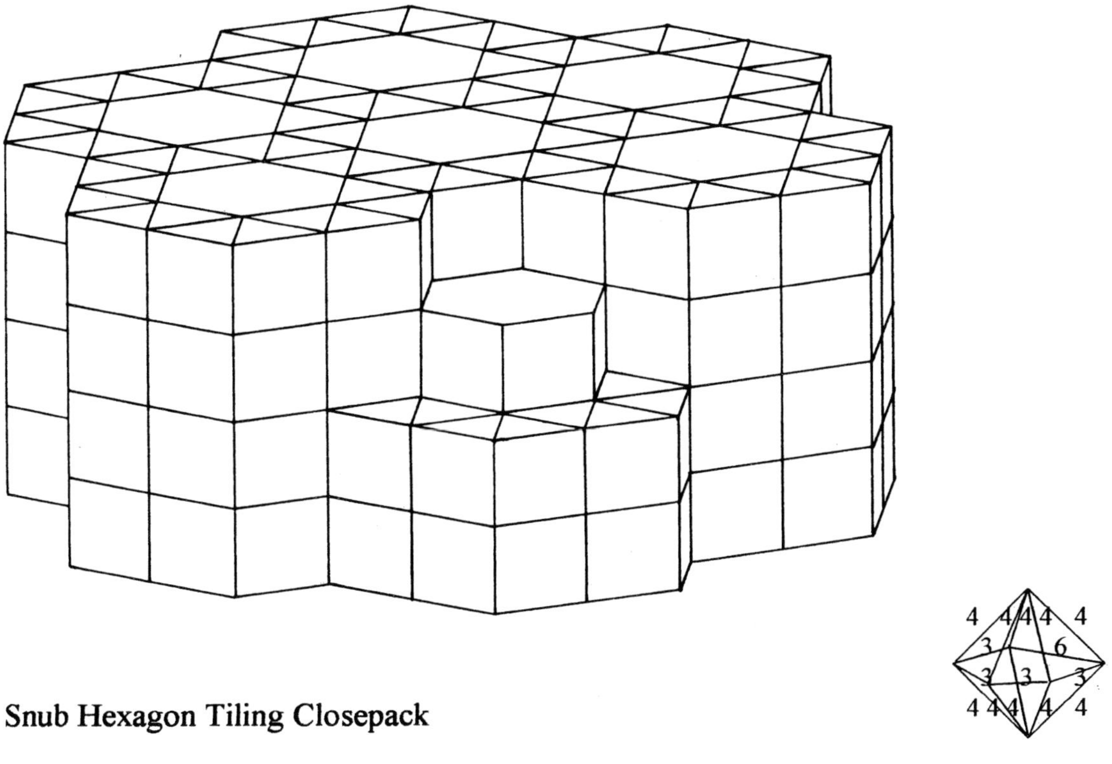

Snub Hexagon Tiling Closepack

Fig.60: Snub Tiling Closepacks

MORE TRUNCATED TILING CLOSEPACKS

Because the centre member of the square tiling truncation series is again a square tiling, there are no further truncations to be found in this family.

However the hexagon-triangle tiling can be truncated to produce a second series with three different types of face present. We thus get two further tiling closepacks with three types of prism apiece, as shown opposite in Figure 62.

Below in Figure 61 is shown the rhombic tiling closepack that occurs at the end of this second tiling truncation series, each rhombi-prism consisting of two triangle prisms.

Obviously we could continue from here to include further tiling closepacks based on the quasitruncated tilings, these involving interpenetrating prisms.

Solids	E/V	F/V	S/V
dodecaprism	5	1 □	2
hexaprism		1 ⬡	2
cube		1 ○	2
		6 □	
Truncated Hexagon-Triangle Tiling Cp			
hexaprism	6	1 △	2
triprism		2 □	2
cube		1 ⬡	4
		8 □	
Rhombic Hexagon-Triangle Tiling Cp			
rhombiprism	5	3 ◇	6
(two vertex types)		6 □	
rhombiprism	8	6 ◇	12
		12 □	
Rhombic Tiling Closepack			

Table 30: Second Series Tiling Closepacks

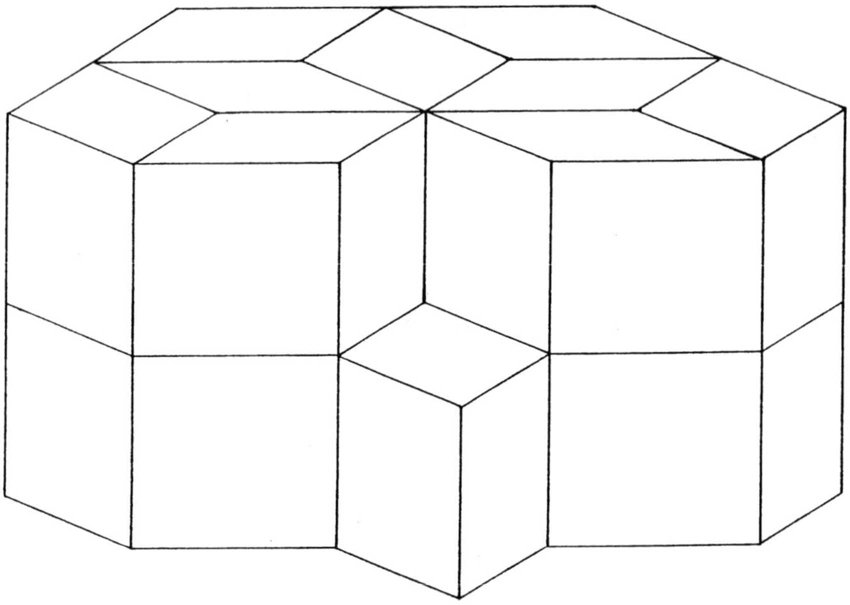

Rhombic Tiling Closepack

Fig.61: Rhombic Closepack

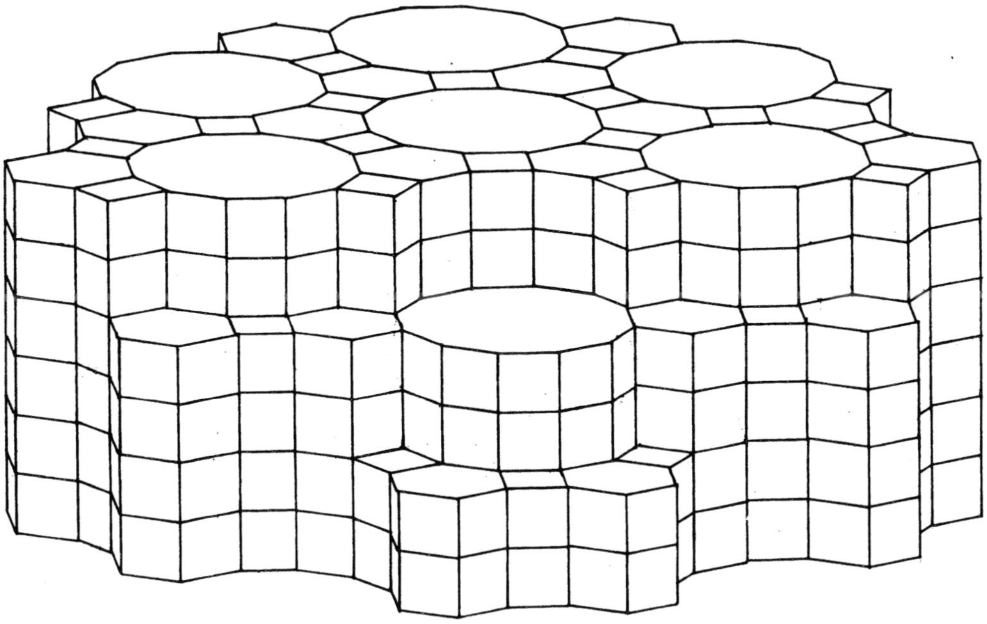

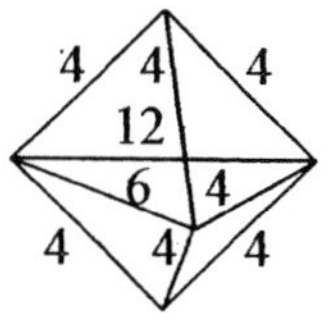

Truncated Hexagon-Triangle Tiling Closepack

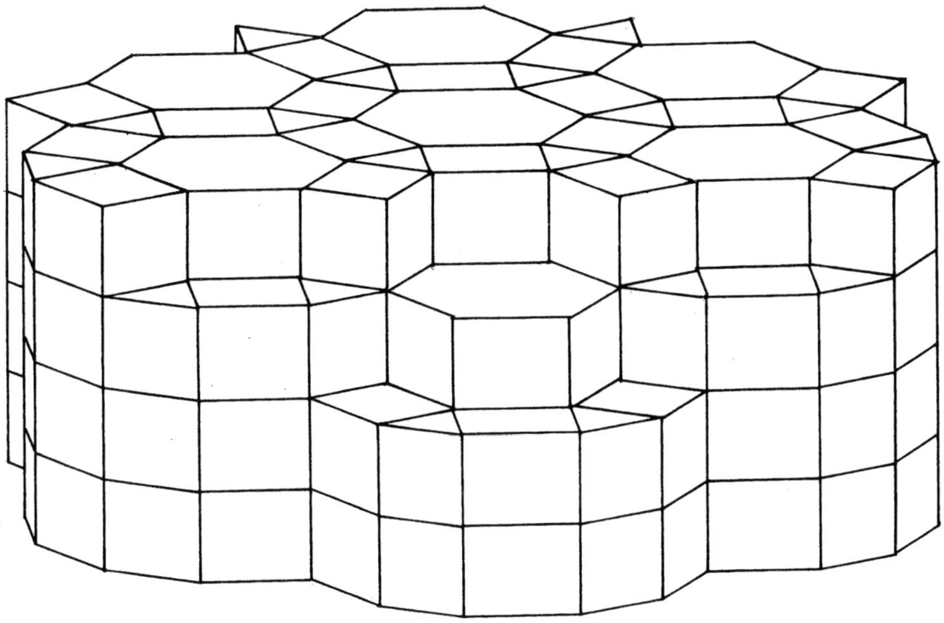

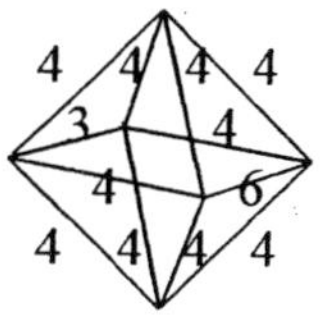

Rhombic Hexagon-Triangle Tiling Closepack

Fig.62: More Truncated Tiling Closepacks

SO WHAT FOLLOWS?

Following on from the polygons, tilings and polyhedra, the 'closepacks and quasi-closepacks' we have examined in this volume of 'The Complete? Polyhedra' have started to go beyond the realms of easy illustration or even easy visualisation.

We seem to have painted a complete? picture of the closepacks and their allies, and although difficult to prove, there should be nothing further to find.

Whilst we have had a hint in the preceding pages that there are further polytopes at higher levels, i.e. in four, five or even more dimensions, it will get increasingly difficult to present these in the visual manner adopted so far in this series. This therefore seems a convenient place to stop our exploration, leaving the higher dimensions firmly in the hands of the algebraist.

We will however persevere. The next logical subject to tackle will be the hypersolids, formed by taking closepacks and putting fewer solids around any edge, thus allowing the figure to fold up into four dimensions and close. This is analogous to taking the tilings and omitting faces so that they fold up into the polyhedra, effectively 'tilings of the sphere'. The hypersolids are thus 'closepacks of the hypersphere'.

If we adopt the policy of only illustrating a small portion of any particular hypersolid, that centred on a particular vertex, edge or solid, we can begin to visualise what is going on in each. The sixth volume in the series will thus tackle these hypersolids and investigate their 'truncations', as we did with the closepacks, followed by the results of quasitruncation, which might get a little difficult.

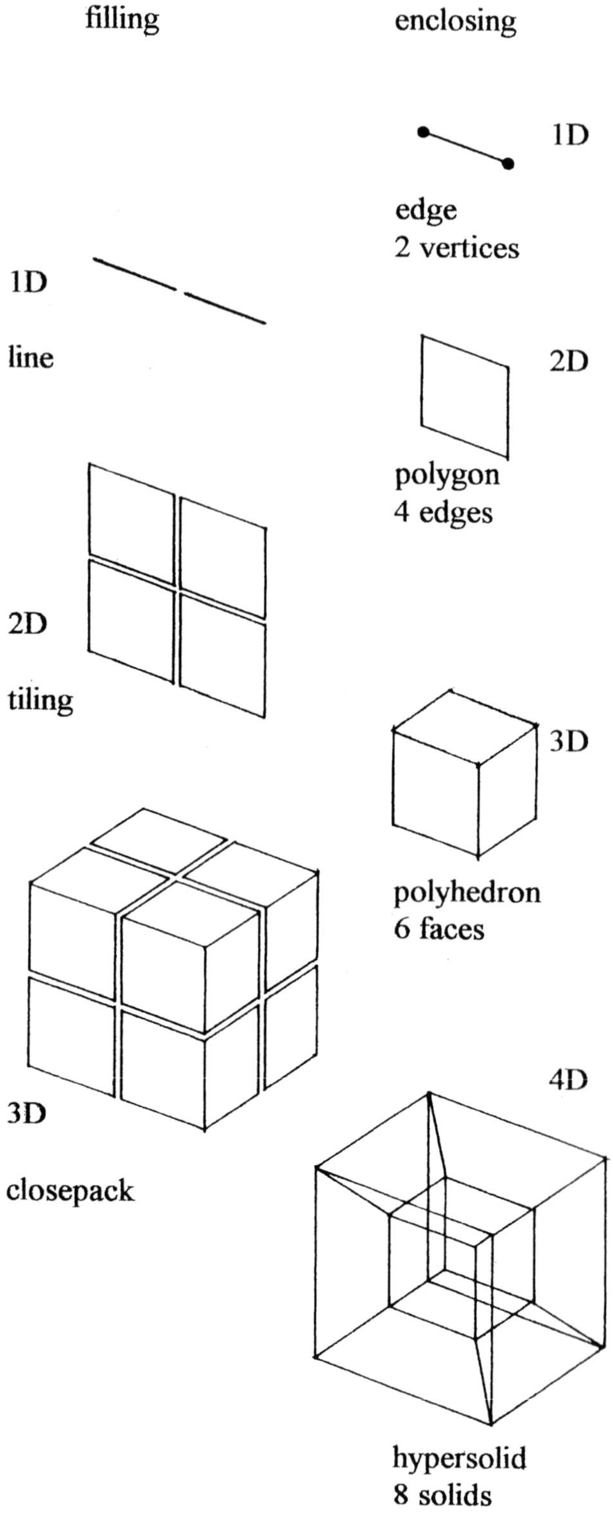

Fig.63: So What Follows?

BIBLIOGRAPHY

Abbott, E A 1962
Flatland
Blackwell

Aste, T & Weaire, D 2000
The Pursuit of Perfect Packing
Institute of Physics

Coxeter, Longuet-Higgins & Miller 1953
Uniform Polyhedra
Phil. Trans. R. Soc. Lond. A246, p401

Coxeter, H S M 1963
Regular Polytopes
Macmillan

Coxeter, H S M 1991
Regular Complex Polytopes
Cambridge UP

Critchlow, K 1969
Order in Space
Thames & Hudson

Cromwell, P R 1997
Polyhedra
Cambridge UP

Grünbaum, B & Shephard, G C 1989
Tilings and Patterns: An Introduction
Freeman

Hargittai, I & M 1994
Symmetry
Shelter

Holden, A 1971
Shapes, Space and Symmetry
Columbia UP

Kappraff, J 1990
Connections
McGraw-Hill

Lakatos, I 1976
Proofs and Refutations
Cambridge UP

Skilling, J 1974
The Complete Set of Uniform Polyhedra
Phil. Trans. R. Soc. Lond. A278, p111

Taylor, P 1990
The Geometry Trip
Nattygrafix

Taylor, P 1995
Additions to the Uniform Polyhedra
Nattygrafix

Taylor, P 1997
The Complete? Polygon
Nattygrafix

Taylor, P 1998
Incomplete Tilings
Nattygrafix

Taylor, P 1999
The Simpler? Polyhedra
Nattygrafix

Taylor, P 2000
The Star & Cross Polyhedra
Nattygrafix

Wenninger, M J 1971
Polyhedron Models
Cambridge UP

Weyl, H 1952
Symmetry
Princeton UP

Williams, R 1972
Natural Structure
Constable

ALSO PUBLISHED BY NATTYGRAFIX:

THE SIMPLER? POLYHEDRA
by Patrick Taylor 1999

being the third part of several comprising the complete? polyhedra

presents the simpler polyhedra derivable from the platonic solids; their truncations, quasitruncations, snub forms and the simpler sub-families and hemihedra, along with an appendix describing the prismatic dihedral forms.

80pp A4 paperback including 62 line drawings and 29 tables

ISBN 0 9516701 4 X £6.00

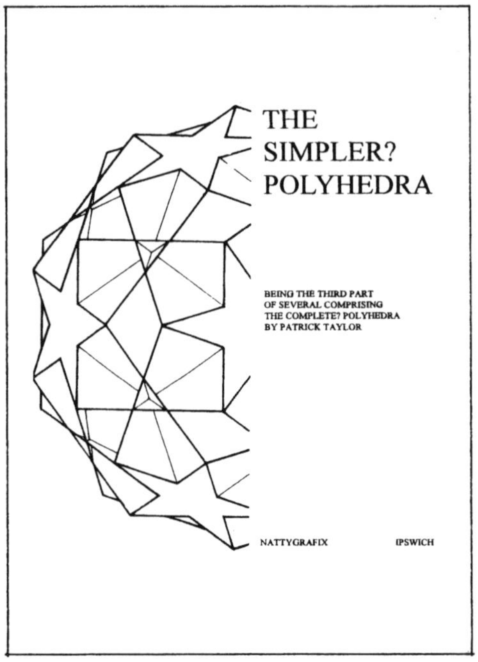

THE STAR & CROSS POLYHEDRA
by Patrick Taylor 2000

being the fourth part of several comprising the complete? polyhedra

presents the more complex polyhedra with star and cross polygon faces; their truncations, quasitruncations, snub forms plus sub-families and hemihedra, along with an appendix describing the prismatic star dihedral forms.

80pp A4 paperback including 71 line drawings and 37 tables

ISBN 0 9516701 5 8 £6.00

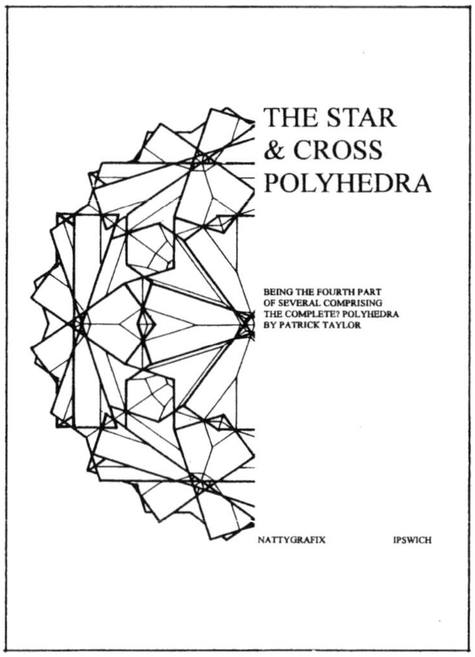